Design and the Creative Process

Design and the Creative Process

Daryl Joseph Moore

THOMSON

DELMAR LEARNING

Australia Brazil Canada Mexico Singapore Spain United Kingdom United States

THOMSON

DELMAR LEARNING

Design and the Creative Process
Daryl Joseph Moore

**Vice President, Technology
and Trades ABU:**
David Garza

Director of Learning Solutions:
Sandy Clark

Managing Editor:
Larry Main

Acquisitions Editor:
James Gish

Product Manager:
Jaimie Weiss

Marketing Director:
Deborah S. Yarnell

Marketing Manager:
Penelope Crosby

Director of Production:
Patty Stephan

Production Manager:
Andrew Crouth

Content Project Manager:
Andrea Majot

Technology Project Manager:
Linda Verde

Editorial Assistant:
Niamh Matthews

Cover Image:
Christopher J. Navetta

COPYRIGHT © 2007 Thomson Delmar Learning, a division of Thomson Learning Inc. All rights reserved. The Thomson Learning Inc. logo is a registered trademark used herein under license.

Printed in the United States of America
1 2 3 4 5 XX 08 07 06

For more information contact
Thomson Delmar Learning
Executive Woods
5 Maxwell Drive, PO Box 8007,
Clifton Park, NY 12065-8007
Or find us on the World Wide Web at
www.delmarlearning.com

Library of Congress Cataloging-in-Publication Data
Moore, Daryl Joseph.
  Design and the creative process/ Daryl Joseph Moore.
    p. cm.
  Includes bibliographical references and index.
  ISBN 1-4018-6164-4
  1. Design, Industrial. 2. Creative ability. I. Title.
  TS171.M654 2006
  658.5'752--dc22
                              2006036403

NOTICE TO THE READER

Table of Contents

Foreword by William Menking

William Menking, Founder and Editor: *The Architect's Newspaper*

William Menking is an architectural historian, writer, critic, and curator of architecture, urbanism, and industrial design. He is a tenured professor of architecture, urbanism, and city planning at Pratt Institute and has lectured and taught at schools in the United States and Europe. He wrote a monthly column "Letter From America" for *Building Design,* the United Kingdom model for *The Architect's Newspaper,* for six years and has been published in numerous architectural publications, edited anthologies, and museum catalogues. He has curated and organized international exhibitions on the British visionary architects *Archigram,* Italian urbanscape photography, the most important collection of post-war experimental architecture from the collection of the FRAC Museum in Orleans, France, the Italian radical architects Superstudio, and contemporary English design sponsored by the British Council on the Arts.

The modern history of "design," both the look of manufactured objects and the process of creating them, parallels the rise of manufactured mass-produced goods. This is most often traced back to the eighteenth century to figures like Josiah Wedgewood or Thomas Chippendale. But the notion of a design process that students can follow to an independent career creating commercial products is not so old and probably dates back to the English arts and crafts inspired schools established in the wake of the 1851 Crystal Palace Exposition and the writings of John Ruskin. In the twentieth century the experiments of the German Werkbund and the pedagogical program of the Weimar Bauhaus were inspirational in creating the early design programs at Cranbrook Academy of Art in Michigan and Black Mountain School of the Arts.

In the United States a few figures like Norman Bel Geddes and then Henry Dreyfus became celebrity designers in the mid-twentieth century, but not until Cranbrook created a school of modern design in the late 1930s and 1940s did the extraordinary possibilities of design become apparent to American industry. The Bloomfield Hills Academy produced many figures who not only designed landmark object and products but also influenced subsequent generations of designers and manufacturers. These included designers like Harry Bertoia, Charles and Ray Eames, Ralph Rapson, Florence Knoll, and of course, Eero Sarineen.

In America today the public recognition of design and the designer's role in the creation of new products has never been higher or their importance to the economy more important. But sadly, design writing—both academic and journalistic—still largely depicts object creation as an act of individual genius springing from the mind of individual creators. This is still true despite all evidence to the contrary that design is a collaborative practice between individuals in an office with fabricators, corporate financiers, and marketing figures to mention only a few of those who come in between design and the realization of a product. The Eames' low-cost wooden furniture, for example, not only evolved out of a group studio, but was a product of their earlier bentwood experiments with wood splints and gliders with the U. S. Navy. In addition, Florence Knoll's classic modern furniture design was a product of the large company that she founded with her husband to market modern office furniture.

The inability to see design as a collaborative practice no doubt comes from the notion that industrial design is an art and even the art of the twentieth century and the fact that many writers and scholars are educated as art historians and critics. But *Design and the Creative Process,* Daryl Joseph Moore's extraordinarily insightful text on American and international design practice, never makes the mistake of focusing too intently on creative genius. The book begins where most design writing never ventures—beyond the personality of the signature designer. It focuses on the collaborative process itself and how designers work through and with their colleagues and industry. Finally, Daryl recognizes that design is not only collaborative in creation and use but also in its reception by consumers—whether it be a toaster, text on a screen, a roadside billboard, or a virtual graphic symbol. It is an important book not just for the profiles of important contemporary star designers and their working methods but as a new way of thinking and writing about design.

Preface

Intended Audience

Design and the Creative Process addresses areas of professional practice that are both distinct and far reaching. The relationship of design, the professionals working at the top of the industry, as well as the societies and people they design these products, communications, services, and environments for are reviewed from the unique historical perspective of each designer. The book provides practical as well as intimate reflections on the approaches to design practice from award-winning designers across disciplines.

The book provides an excellent resource and reference tool for both intermediate and advance level design students in communications design environments on the undergraduate and graduate level (graphic design, advertising design, package design, motion/time-based graphics, industrial design, environmental design, architecture, fashion design, etc.). It is also a critical resource for design professionals and marketers who want a straightforward and concise look at the contemporary professional practices of a diverse group of international designers, working in multidisciplinary environments.

Design and the Creative Process provides readers with an analysis of design solutions arrived at, unlike the usual examples often found in publications with short reaching goals, merely reprinting the work with little background information on the examples presented or the professional practice of the designer. The critical analysis of each example in *Design and the Creative Process* will inform those creative problem solvers with a vested interest in the significance of design to their businesses as well as inform readers with a general understanding of design and its relationship to society.

The unique approach of each designer highlighted between these pages provides readers with clear evidence of that creative process, enabling each example to be a marker of success and a road map to inspiration.

Background of this Text

The text for this book was developed over a three-year period through onsite personal interviews of the designers highlighted in the book. The goal was to develop an informative, personal text that would enlighten its readers with insight previously unavailable as a reference on the specific nature of design and its professional practice as illuminated through the experiences of these award-winning international designers.

Through the qualitative analysis of each compelling story, *Design and the Creative Process* gives its readers an inside look at the professional practices of contemporary designers that would otherwise be unavailable. This is the information void that *Design and the Creative Process* begins to fill. As a teaching tool, the book will give design faculty the language needed to help aspiring designers develop the processes and analytical skills required in formulating solutions for communications, products, and services that are a part of our everyday lives. Historically, design has the power to effect and arguably continue to shape the societies in which we live, work, and play in an increasingly profound way. Students will gain an understanding and appreciation of this creative process through the behind-the-scenes review of each designer's background and philosophy across disciplines. Each designer can be reviewed in the classroom or studio during the course of a semester, critically focusing on specifics in their process such as relevancy and need driven by client, consumer and marketplace issue, and function and rationale. This provides the instructor with contemporary examples that can be used as a springboard for advanced level project execution.

About the Author

Born and raised in Newark, New Jersey, Daryl Moore graduated from America's first visual and performing arts high school, Arts High located in the Garden State's largest city. He holds an MFA degree in advertising design from Syracuse University, and a BFA degree in communication design from Pratt Institute.

He is an elected Fellow of the Royal Society for the Encouragement of Arts, Manufactures, and Commerce (RSA), London, England, as well as the department chair of the Art and Design division of the College of the Arts at Montclair State University, New Jersey's second largest and fastest growing comprehensive university. Mr. Moore's professional background spans two decades and encompasses illustration, creative direction in advertising, and senior art direction and design in corporate and graphic design environments. He is a former founding partner of the design and advertising firm Visual Communications Group (VCG), located in Princeton, New Jersey.

His multidisciplinary design practice Simpatico serves organizations in the area of design, advertising, promotion, and product and brand development.

ISBN: 1-4018-6164-4

Acknowledgments

I would like to thank the following list of individuals, without whose support, encouragement, and generosity over the years this book would not be possible: Jim Gish, Jaimie Weiss, Andrea Majot, Marissa Maiella, and Niamh Matthews of Thomson Delmar Learning, Deepa Gosh of ITC, Paula Scher, Alan Chan, Kyle Cooper, Deborah Sussman, Mi Qiu, Patrick Robinson, Alex Lee, Stefan Sagmeister, Ernest Lupinacci, Bill Menking, Phyllis Weisband Fibus of Prologue, Stephanie Karlis and Gretchen Holt of OXO, Angela Killoren of Sussman Prejza,

Naoko Ito, Kate Etter of Puig Fashion, Selina and Ida of Alan Chan Design, Kurt Koepfle and Audrey Christensen of Pentagram, Kelly Parrotto, Gary Grey, B. Martin Pederson, Dr. Bakke SU ISDP, David J. Passalacqua, Alex Bostic, Rudy Gutierrez, Liang Tian and the entire MQ Modern Art Workshop staff in Shanghai, Han Jaiying, Sergio Young, Harry Massey, Gregory Steven Kelly, Derrické Summerfield Florence and Derrické R. Summerfield, Andrew Teheran, William Moore, Kevin Moore, Amiri Baraka, Joe White, Clarence White, John Sears, Alberto Baptiste, Tommy Lyman, Raymond Gregory & SMB et al, Darlene Susco, Kristi Shuey, Roger White and Audria Nunnally of VCG, Larry & Sheila Miller, Marc Connelly, Audrey Stone, Luis Spindler and Hilton J. Otero of Arts HS, Ms. Kae. Payne and Mr. Jeffrey Pollack of Maple Avenue Elementary School, Rich Carrara, Iris Cruz, Russell A. Murray, Roger Bruan, John Luttropp, Nancy Goldring, Lynda Hong, John Czerkowicz, Peter Barnet, Bob Browning, Dorothy Heard, Martin Greenwald, J. Catherine Bebout, Mimi Weinberg, Klaus Schnitzer, Bill McCreath, Patricia Lay, Winfield Parsons, Scott Gordley, Julie Heffernan, Eileen Foti, Tony Pemberton, Bill Jennings, Anne Betty Weinshenker, Susi Colin, Elizabeth Valdez del Alamo, Sabine Eck, Walter Swales, Chris Schade, Josh Jordan, Denis Feigler, Linda Reilly, Elaine Flint, Cathy Le Claire-Wright, Janet Filomino, Karen Guancione, Pamela Scheinman, Diane McNamee, Andrew Atkinson, Ruth Rendleman, Marge Joyce, Charlene Mickle, Louise Davies, Tara Thurber, Hisayo Kushida, Ruijun Shen, the students of Montclair State University's department of Art & Design, Dr. Geoffrey W. Newman, Dr. Ronald Sharps, Linda Davidson, M. Teresa Rodriguez, Carlos Ortiz, Dr. Richard Lynde, President Susan A. Cole, Marina Cunningham, Dazhou Wang and the faculty, staff and students of ECNU School of Design P.R.C., Dr. J. Harry Smith & family, Ty Moore, Monique Moore, Anthony Ricardo Moore, my parents Richard and Valeen Moore, my children Tate, Kai, Blair, and Ashley Moore and of course Naoko Maekawa Moore my partner and compass in life.

Thomson Delmar Learning and the author would also like to thank the following reviewer for her valuable suggestions and expertise:

Susanne Manheimer
Chair, Graphic Design Department
The Art Institute of California – Los Angeles
Santa Monica, California

Questions and Feedback

Thomson Delmar Learning and the author welcome your questions and feedback. If you have suggestions that you think others would benefit from, please let us know and we will try to include them in the next edition.

To send us your questions and/or feedback, you can contact the publisher at:

Thomson Delmar Learning
Executive Woods
5 Maxwell Drive
Clifton Park, NY 12065
Attn: Media Arts & Design Team
800-998-7498

Or the author:
Daryl Joseph Moore FRSA
P.O. Box 43610
Upper Montclair, New Jersey
moored@mail.montclair.edu
daryl@simpatico56.com

Dedication

Design and the Creative Process is dedicated to the continued celebration and promotion of design, its critical link to the fulfillment and growth of our society, and to the educators and professionals who hold onto these ideals while never losing sight of the power of design to change and improve the daily lives of us all.

Finally, this book is dedicated to the Designers who so graciously shared glimpses of their lives in design with me, which give these pages life and continued hope for the transformative power of design toward the creation of a better world.

DJM

Introduction

Design And The Creative Process

Design and the Creative Process is a journey and exploration into the intimate world of eight internationally known designers and their respective design companies. The group includes names that are familiar, as many of the designers selected for this publication have been working successfully for several decades. Others may be less familiar in that their playing fields are as far away as Hong Kong and Shanghai, China. That the text begins with Paula Scher, who is on any "informed" short list of relevant American designers over that last twenty-plus-years—is appropriate and fortuitous for the reader and author. I am indebted to her, as I am to all the participating designers, not just for the time she so graciously spent with me in review of her ideas on design—specifically her identity development for the Citigroup—but also for her example, which set the tone and pace for the other participants who followed. She has set a high mark, as she continues to be a major force in design and all that is implied by it, as our society continues to evolve driven by a culture that stopped being local with the creation of "MTV."

Paula Scher is in very good company here, as her story is just the tip of the creative iceberg and is followed by the prolific Alan Chan, principal of The Alan Chan Design Company based in Hong Kong. Chan's body of work rivals any produced in the world—AIGA medalist, B. Martin Pederson, owner, publisher, and creative director of Graphis Inc. got it right when he suggested that Chan's Studio was one of the top design studios in the world. My brief connection with Pederson, an important figure in design in his own right, was during the completion of the graduate degree I pursued at Syracuse University in the early 1990s. He spoke to

P. Scher

me then about the importance of realizing that design should be the starting point—not a place to get too comfortable. His suggestion was that it was in fact a vehicle to be driven to the next place—a means to an end—much like the Porsche BB I had gracing the cover of the magazine design project I submitted for his review for the course (I got an A). I mention this not as a clumsy and smug way to promote my graduate school experience and drop the name of Pederson, but in speaking particularly about the level of respect from leading design practitioners that Alan Chan has garnered over time. He has embraced this idea whole-heartedly through the transcendence of his work. His amazing

understanding of design and high aesthetic comes from someone who has had the equivalent of "ten minutes of formal design education." Yes, he is very special and also committed to excellence in design. His intuitive approach and embrace of multiple cultures is reflective of his upbringing in Hong Kong, a city where Eastern and Western thought have been in the same "bed" for decades, harmoniously so in the mind and visualizations of Chan. And like many who reach the highest level of practice in this very competitive and critically important profession, Chan has used design as a wonderful vessel allowing him to cross significant boundaries, including graphic and

A. Chan

had the opportunity to embrace Western modernism as well as one of the first contemporary artists exhibiting outside of the country in the early 1990s.

His multidimensional visual arts background includes printmaking, painting, sculpture, urban planning and design, architecture, environmental, and interior design. I found it fascinating and wondered how one individual could position himself to be successful in so many areas of design. Success at the level that he has achieved in any single one of those areas would be enough for many of us. But as you will find on the following pages, this grandson of a shaman, is unique and always in search of a higher plan for the people of China. He seeks to keep the traditions of the past intrinsic to the ideals of today while at the same time designing and creating the structural and visual spaces in the built environment of China's bright future.

Mi Qiu

advertising design, interior design, product design, and more recently his exploration into the world of installation and exhibition as a fine artist.

I met Chan on a chance encounter during a visit to Shanghai, China, in 2003. Mi Qiu, Principal of MQ Modern Art Workshop & Environmental Design (based in Shanghai), suggested that I take a short walk with him after dinner to the adjoining restaurant to meet a friend of his from Hong Kong. That friend was Alan Chan. That these two very important designers were close friends was more than I could have ever hoped for. Mi Qiu's story, which is included here, is perhaps the least known of the designers highlighted on these pages (in America, that is). His story is compelling on so many levels. A survivor of Mao's Cultural Revolution from 1966 to 1976, he is a member of an important group—that of the first generation of Chinese artists who

Moving to time-based design, specifically motion graphics for feature films and television, Kyle Cooper of Prologue Films is highlighted as well. With his creation of the opening live-action title sequence design for the David Fincher psychological thriller Seven, he confirmed the opinion amongst critics and the movie-going public that not only was he heir to the vacated throne of the late great Saul Bass, who historically garnered the most recognition but was not alone in the innovative use of motion graphics in title sequence design — but also that the creation of title credits for feature films is a critically important area of design practice. Bass was joined by the likes of Robert Brownjohn and Maurice Binder (best remembered for their provocative James Bond credit sequences) and others in the movie title design canon. This design element can compel an audience to excitement, or in the case of the Fincher movie—abject fear and suspense.

More recently, Cooper has helped the American movie-going public enjoy the world's favorite web-slinging superhero Spider-Man, as he created the arresting title sequence designs for the original film and for its record-breaking sequel. Cooper's story is as compelling as the other designers; he integrates experiences from the modernist and post-modernist side of his development as a designer working in a time-based medium.

As a Yale graduate student, Cooper studied under the late Paul Rand, and considers Rand to be the critical mentor in his creative development as a designer. Cooper's West Coast studio—perched above the stunning shores of Malibu, California—is about an hour's drive from the Culver City based Environmental Graphic Design firm of Deborah Sussman of Sussman Prejza.

Sussman is a "grand dame" in the world of environmental graphics and design. The road that lead to her present position in the industry winds through the famed Charles and Ray Eames studios as well as an eclectic education that included studies with innovative choreographer Merce Cunningham and musical composer John Cage at Black Mountain College. These

D. Sussman

K. Cooper

S. Sagmeister

philosopher Edward de Bono, Stefan's creative process is his ability to think in creative ways. He dismisses the idea of thinking outside of the box—there is no box. His ground-breaking and innovative use of compact disc (CD) packaging for the music industry shows his ability to adjust and utilize technological changes—in this case the move from the standard record packaging that predates CD packaging. In this he is taking charge of change rather than be limited by it.

Sagmeister's propensity for in-your-face design compels him to take a very personal approach in arriving at the right solution even if it means disrobing and having type "set" literally on his upper torso via razor blade cuts. His approach to design is noble and sincere, and his generosity and insight bring this text to life.

Alex Lee, a graduate of Harvard and Parsons, is the President of the OXO Company located in New York City. The entrepreneur Sam Farber, who "just wanted to make everyday kitchenware that was durable, easy to use and good to look at," founded the company in 1990 and personally

varied experiences are defining elements in her approach to environmental graphic design and have propelled her and the firm she co-founded to the top of the profession.

Her seminal work produced for the 1984 Olympics put the studio on the international map and was heralded as one of the most important designs of that decade. Her creative process takes an anthropological approach to design in the built environment, where she thoroughly seeks out "the DNA of the project." Her generosity of spirit and willingness to give of her time validate the importance of this book as more than a celebration and appreciation of design, rather, it is an inside look into the choices made by individuals, such as Sussman, who are the innovators in the field.

Stefan Sagmeister of Sagmeister, Inc., is an innovative and conceptual designer. His design studio, situated in the center of New York City, has produced provocative and relevant work for the music industry and for corporate entities for more than a decade. A confirmed disciple of Tibor Kalman and

A. Lee

Sam Farber

hand picked Lee as his successor. The OXO story and the company's embrace of universal design have helped it maintain a winning formula in the world of product design. OXO's success in business and manufacturing is based on the Harvard business school model. However, it is the unique sensibilities of the people leading OXO who have made the company a virtual hallmark for smart design. — Their stories will motivate readers and those in design practice and education who truly appreciate smartly designed products that are affordable, transgenerational, and durable enough for today's demanding households.

Rounding out the group of highlighted designers is fashion industry's Patrick Robinson, who seems to have the market on the reinvention and rejuvenation of quiescent, but historical, fashion houses. The list is quite impressive: former design director of "White Label" for Armani (Milan, Italy); former creative director and senior vice president of Anne Klein (New York); former designer of his own line, Patrick Robinson Inc. (New York); former creative director, Perry Ellis Womenswear (New York); current creative director for Paco Rabanne (Paris, France). Upon meeting Mr. Robinson in August 2004 to discuss his inclusion in the book, I was amazed at his accomplishments starting with his position in Milan for Armani when he was only 24. This was just prior to the launch of the Spring Collection for the Perry Ellis Collection that took place during Fashion Week in New York City; this was his final body of work for that famed fashion house. He is continually reinventing himself but his unique approach and creative process remain intact.

This brief introduction rounds out the talented and generous group of some of today's most relevant and successful contemporary designers. They are as diverse as they are gifted—working on an international stage in the management and

P. Robinson

creation of products and visual communications — from tearooms and teakettles, womenswear to corporate identity, and environmental design, both graphic and architectural.

The designers included are the main players who provide frank and informed thinking on design and their creative process. They have opened their doors and minds because of their belief in the critical importance of design—not only as it affects consumers in the marketplace, but its importance to society both globally and locally.

As an educational tool this book provides readers with an intimate review and analysis of the experiences and professional practice of eight contemporary designers; their work crosses disciplines and continents. Dipak Jain, Dean of Northwestern University's Kellogg School of Management stated in an interview with @Issue, a business publication produced by the Design Foundation in Massachusetts

> . . . competition today comes from everywhere (globalization and technology). Given these choices consumers [and] end users are more demanding and better informed than ever. More companies are using design to differentiate their products and services—by design— not just the aesthetics of the product, but the total customer experience.

His comments underscore the importance of design in a global way. The idea of moving from "high tech to high touch" in relating to people and their specific needs is a provocative term, but nevertheless appropriate in describing what good, smart design must focus on if it is to continue to be an agent of progress in the human experience.

Thomas L. Friedman's global perspective on the history of the twenty-first century and the convergence of technology confirms that countries previously excluded from the supply chain of services and manufacturing have now taken a front seat in that process. The importance of good design and its correlation to the success or failure of products in the marketplace has never been more crucial. More significantly, the ability for design to make a difference in the world in which we live is as important as our need for consumer goods and services. We are now designing for the great society that is the new world we all occupy.

Design and the Creative Process is the thread that connects each designer highlighted in this book; it reaches beyond the typical and often visionless texts concerned only with design formulas and the latest display font or creative trend. Each creative thinker and their candid insights, which they have so graciously provided for students, faculty, and professionals, builds on their own perspectives and focuses on the importance of good design and its critical role in society. I hope you enjoy the text.

DARYL JOSEPH MOORE, FRSA

Mergers and Acquisitions: Divergent Corporate Cultures and the Power of Design

Client—Citigroup: Sandy Weill Chairman, Principal Owner (The Travelers)

John Reed, Principal Owner (Citicorp), and Susan Avarde, Director of Global Brand Management

Creative—Pentagram: Paula Scher and Michael Bierut

Figure 1-1.
Paula Scher.

Introduction

The creative process is an exploration of design and of the people integral to it—this process is critical and relates directly to the success of effective design solutions. In this chapter we will review and explore the redesign of the Citibank logo that began in the late 1990s as a result of the acquisition of Citicorp by The Travelers Insurance Group, a firm that has been in existence for over a hundred years. Congress had approved changes in the law that allowed the coexistence of insurance and banking entities under one roof. As a result, the aforementioned acquisition propelled Citigroup to its current position as the world's largest financial services corporation. Both organizations had, and have, a distinct business culture, which made for an uneasy transition. Creating a visually relevant and enduring identity for a company as large as Citigroup would be difficult under the best conditions. Add to this the politically charged environment of the acquisition, employee concerns about job security, and divergent perspectives from the top of both organizations. This scenario had all the makings of a potentially bad client experience with the risk of not reaching your expectations creatively. However, with the cooperation of a focused team of professionals at Pentagram, Lippincott Mercer, and Fallon advertising and a progressive group of brand managers on the client side—today the Citi logo is unique in the crowded field of corporate identity. With that brief introduction, we will review the creative process that lead to its development.

There is a distinct process behind all successful design solutions, which is clearly evident with the Citibank logo design. However, depending on the creative agency, the design group, or the individual, that process varies distinctly from job to job. Motivated by the specific needs of the client and the target audience, Paula Scher of Pentagram put this process in motion. This nuanced and detailed ongoing creative process was born out of communication with all participants and critical analysis of all relevant issues.

Overview

Scher brought over three decades of creative problem solving to the enormous task of developing a new identity for the world's largest financial services corporation, Citigroup. Her wealth of experience and creative talent enveloped a shrewd and thoughtful business acumen that clearly resonated with the client and her creative associates at Pentagram, Lippincott Mercer, and Fallon. Scher blazed a creative path that began professionally as a designer and art director with Atlantic and CBS records. Her progressive album cover designs in the 1970s were the foundation for co-founding her own design firm in 1984, Koppel & Scher. The design firm existed successfully for eight years. In 1991 she joined Pentagram as a partner. In reviewing her work, it becomes clear that Scher is a giant in the male-dominated world of graphic design. From the typographically wonderful LP covers for recording artists like The Yardbirds and Phoebe Snow to the amazing identity program of the Public Theatre, the breadth and depth of her work speaks to her

talent as a conceptual and articulate visual communicator armed with a very practical approach to getting to the essence of the creative problem.

On corporate strategy and culture Scher states . . .

> Consumers don't know, and do not want to know the inner workings of the company's services they need. However, it is important as a designer working for the company to have this information—the

Figure 1-2.
Public Theatre Poster | Three, one-act plays written by Steve Martin. (See also Figure (ii) in the color insert.)

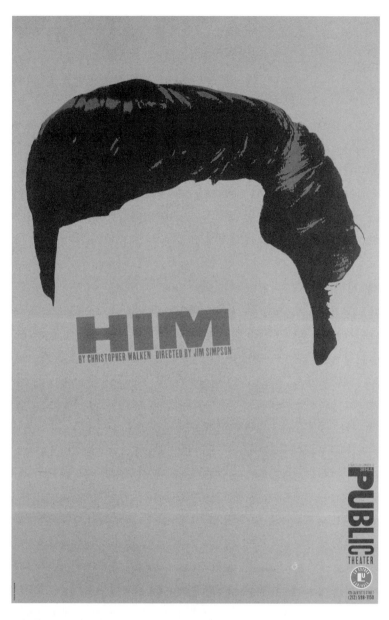

Figure 1-3.
Public Theatre Poster |
A play by Christopher
Walken, directed by
Jim Simpson. (See also
Figure (iii) in the color
insert.)

consumer just wants to know that it will work when they use it, whatever it is.

They want to be able to get to those services easily.

She has design work in permanent collections of the Museum of Modern Art, the Cooper-Hewitt, and the National Design Museum in New York City. Her work is also exhibited internationally at the Poster Museum in Zurich and the Centre Georges Pompidou in Paris.

Her talent and sagaciousness—coupled with her remarkable ability as a designer—

enabled her to successfully navigate the distinctly divergent corporate cultures of the Travelers Insurance Group and Citibank.

When Worlds Collide

I am reminded of the Seinfeld episode in which the character George Castanza is totally put off and bordering on hysteria when he realizes the dynamic of their inner circle of associates is about to change. "When worlds collide" he noted in his usual desperate tone, "crazy and unexpected things can happen"—usually bad things. Sometimes out of the most fantastic and unforeseen collisions of creative and divergent corporate forces something good can happen. These next few pages present evidence that suggest the latter from the design perspective, culminating in the successful Travelers and Citibank corporate identity collaboration.

The Creative Process: Key Players

Primary individuals on the client side were Sandy Weill, chairman and principal owner of The Travelers; John Reed, Citicorp principal; Michael Wolff, the brand and identity strategist based in Great Britain (who contacted Pentagram Partner Michael Bierut initially about the project); and Susan Avarde, head of global branding for Citibank. Visually speaking, the respective companies had similar paths. The task of developing an effective and compelling corporate identity program that is consistent with the goals and personality of the company—while remaining visually relevant and memorable to the public—is challenging enough without negotiating competing elements within the organization. The elements I refer to are often based on personality differences and dissimilar points of view, both aesthetically and practically, that are a result of distinctly different cultures.

Figure 1-4.
Public Theatre Poster | A play by Suzan-Lori
Parks, directed by Richard Foreman.

The aforementioned speaks to the client-side perils of a complex organization, often referred to as account conservation from an agency perspective. As a designer, working in today's aggressive climate—a climate of individuality and entrepreneurial spirit where companies, products, services and information are often awash in a vast ocean of visual noise—it is not enough to merely solve the creative problem at hand, utilizing arbitrary forms in space in the construction of visual units to represent them. Designers must reach higher. In his explanation of visual strength, Hans Schleger one of the great Graphic Designers of the twentieth century refers to a Paul Rand quote . . . "Symbols are a duality. They take a meaning from causes good or bad—and they give meaning to causes good or bad . . . . The vitality of a symbol comes from effective dissemination . . . it needs attending to to get attention. The trade mark/logo is not a sign

of quality—it is a sign of the quality . . . ideally trade marks/logos do not illustrate, they indicate. . . ." Schleger further states "The best trade marks are seen and recognized—and not translated. Their impact can be deep, and after a while they become permanently absorbed by the person receiving them. . . ." Added to this is the ideal timelessness of the final logo design.

Figure 1-5.
Public Theatre Poster |
Jennifer Lewis's one-
woman show.

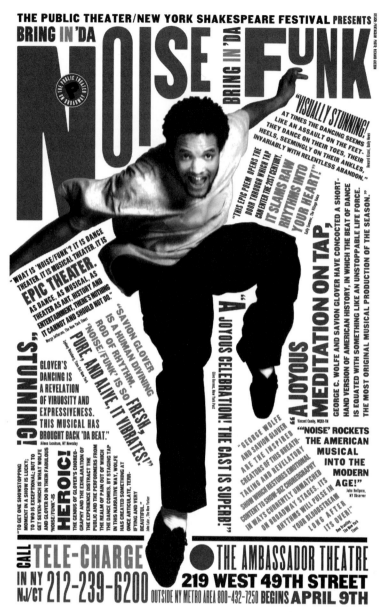

Figure 1-6.
Public Theatre Poster and teaser ad | Bring In 'Da Noise, Bring In 'Da Funk promotional campaign, photo by Richard Avedon.
(See also Figure (iv) in the color insert.)

Designers must have the ability to understand the essence of the statements by both Rand and Schleger—in addition to being skillful and nuanced in steering the powerful and informed individuals who make the final decisions on the client side. To this end the Citi logo was very much "attended to"—by Scher and her creative colleagues. It is noted here again that without the additional focus and creative partnership of the client the success of the project would not have been realized.

Visual Beginnings

Travelers logo had a long and customer-friendly shelf life. Travelers had been insuring individuals and groups since 1864, and the logo featured a very idiosyncratic and scholarly logotype. The company name "Travelers" was set with an initial cap "T" in a tightly kerned bold face bracketed serif font with the familiar red umbrella tucked to the right of the small letter "s". This device provided the visual innuendo implying protection—which draws from the rainy day metaphor that most people could relate to. This device reinforces the company name, while relating it to the product it represents and connecting it to its constituents on a very basic level.

Visual innuendo is the implied reference to the company's core values, and in today's marketing language, its brand essence. This "meaningful" device (visual innuendo as Paul Rand described it) is critical and contributes to the true measure of a logo and its relevance in the consumer marketplace. The creative approach to the new logo pays homage to this ideal. In arriving at the new, simplified logo, Scher admits to conceiving of the new structure just forty-eight hours after meeting with the client. One of the first strategic steps taken in the process was the shortening of the corporate name from Citibank to Citi. This was initially suggested by Michael Wolff and concurrently arrived at by Scher and Michael Bierut of Pentagram. The project called for the creative team to have a completely developed logo that would be ready for launch in ten weeks. As has been the case in the career of Scher, she was once again ahead of the curve. As she explained it, arriving at the final design solution of the logo was not a great stretch for her, in that it was clear that once she selected Interstate as the font of choice, the visual nuance of the small letter "t" took her

Figure 1-7.
Public Theatre
promotional first season
T-shirt. (See also Figure (i)
in the color insert.)

directly to the form of the red umbrella from the Travelers logo. However, Scher's ability to arrive at the appropriate typographical solution quickly is a direct result of her broad knowledge of type—which emphasizes a focused study of its application, form, and function as a means of communication as well as its [type] strength as a design element. This she has consistently achieved throughout her career and professional practice.

Design Simplicity

Similar to great designers who have preceded her, Scher's appreciation for simplicity in design is evident in her final submission for Citi, which she says "is exactly like what I had come up with two days after the initial meeting with the client." "Marks that are more simplified tend to be more successful and visually relevant—something that becomes pure and clear."

She was not suggesting that her ideas were an easy sell to the client; let us not forget that this client was actually two clients. Scher's creative process requires that she understand how the company makes decisions and who the key people are during the review and decision-making process. To achieve this she has become a keen observer of human behavior. She contends that a designer cannot successfully complete the job without knowing the players in the game and their personalities. From this standpoint "knowing the structure of the company is very important." On the one side there was the strong affinity for the red umbrella and its incorporation in the new logo. Then there was the blue-wave group, who proved to be a bit more problematic and required the creation of a new methodology of its use with the new mark.

The analogy that Scher used convincing the client that the new mark was the best solution was that of a red dress.

Most times you are not going to persuade a corporation which has been

Figure 1-8.
Poster design for Paula Scher's AIGA lecture, University of North Carolina at Chapel Hill.

WARNING: Paula Scher in Raleigh
Presented by AIGA, Sponsored by FGI
Thursday, February 24, 1994, 7:30pm
100 Hamilton Hall, University of
North Carolina, Chapel Hill

wearing the same blue dress to the dance to suddenly change and decide to wear a bright red one instead. So what I do is

get them to wear a red belt with the blue dress to the dance.

This is a risk a client is more likely to take. Once again Scher is making design language human—"decoding the process and the problem." She calls it deconstructing the process, and she attributes her realization and gradual acceptance of this through her association and working relationship with Michael Bierut. She felt that she needed to take a more collaborative and less adversarial approach with her clients during the creative process. This ties in directly with the importance of presentation in successful design projects; the use of a simple language in defining the creative direction of an idea provides the client with the tools they need to make the best decision enabling them to be successful in their business.

The Pedagogy of Presentation

The strategy of Pentagram in their presentation to Citigroup was built around the idea of the bank of the future. This

Figure 1-9.
Ballet-Tech Season promotional poster 1997. (See also Figure (vi) in the color insert.)

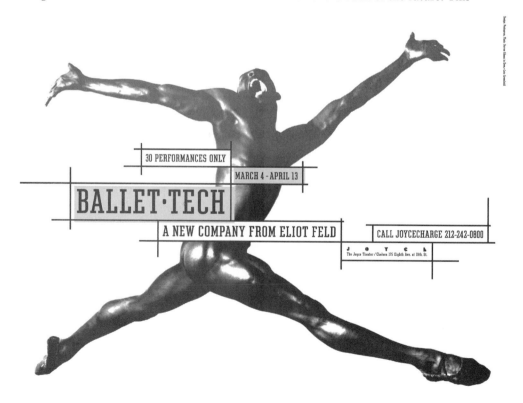

30 PERFORMANCES ONLY

MARCH 4 - APRIL 13

BALLET·TECH

A NEW COMPANY FROM ELIOT FELD

CALL JOYCECHARGE 212-242-0800

J   O   Y   C   E
The Joyce Theater / Chelsea 175 Eighth Ave. at 19th St.

Figure 1-10.
Original Travelers red umbrella logo.

description included a profile of an eleven-year-old girl's idea of banking services, which is very different from our own approach as consumers. She has little need to understand branding issues like blue waves and umbrellas. Her environment is filled with the symbols of the giant brands that permeate the consumer landscape, such as Nike, Coca Cola, and McDonalds. She will probably do most of her banking from wherever she is online; she wants ease of access. Michael Bierut put together a presentation that demonstrated with dramatic clarity the new logo's rightful position among the logos that are part of our visual culture. The presentation validated the visual strength and relevancy of the new logo among the field of successful brand identities, which were placed side-by-side on the presentation board for client approval. It was as if the mark had been around for several years.

The Anatomy of an Idyllic Logo

It is not often that a logo design is weighted with all of the expectations of the newly combined Travelers and Citibank corporation. The logo must exhibit the positive visual qualities needed to reinforce the corporate vision of the people and the products it represents. The Citi logo went beyond that, and is one of the most

beautifully realized marks on the recent corporate horizon. Scher's design is as structurally sound as an architect's rendering of the New York City skyline, and when applied environmentally, it enhances the space it occupies. This is the result of spatial considerations that were strategically made by the designer in executing the final logo design. The precise open-spaced kerning used in Scher's optically adjusted logotype makes the unit structure a testament to strong composition and the importance of legibility and visual integrity. This should be one of the innate qualities of successful designers that Scher spoke passionately about during our interview. She is not inclined to define a specific set of rules or process that she employs when beginning a job, and I suspect that this is precisely her creative process. Her experience and broad esthetic are the backdrop for a canvas that incorporates her time-tested and conceptually fresh approach to design. Her designs are successfully transgenerational. She works from a very informed perspective paying homage to history and current cultural perspectives. She has also been influenced by architecture, and this has played a major role in her approach to design. These creative elements come from years of professional practice, and in some instances, can be traced directly back to her more successful album jackets. In fact as illustrated in Figure 1-4 and Figure 1-5, we can draw a direct visual parallel to the progressive, architecturally unique New York City skyline and the consistent stroke weights of the arch (umbrella in the abstract) and its relationship to the letterforms. The implicit vertical stroke weights of the font Interstate, protected by the red arch, relates not only to the original red umbrella but also to other positive

symbols such as halos and rainbows. Such is the power of positive visual innuendo. The fact that the red arch is a more abstract form with several visual connotations makes it all the more powerful to the viewer.

The red arch is now the visual beacon of the world's largest financial entity calling out to its loyal constituents with confidence and integrity. As validated during the initial presentation, it feels as though it has been a part of the visual corporate landscape for years. Scher noted that today (partially due to the rapid employee turnover) there are few employees who remember the old mark.

"It's as if this was always the logo," she states. One carryover from the old mark that had to be addressed was the blue wave background, which those on the Citibank side of the business insisted remain a part of the new mark. Scher's implementation of the blue wave demonstrated the multiplicity of usage that the wave could have on all elements where the logo could be applied—from environmental signage to individual credit cards. Anatomically the new logo satisfied both the old guard and the new through the use of design and color—the red of the Travelers original umbrella logo and the sans serif font and blue wave of Citicorp. Not an easy recipe for design success where generally speaking less is more. The Citibank identity project as described by Scher took a reasonable and logical approach to solving the problem. Given the judicious

Figure 1-12.
Citi logo thumbnail sketch.

support of the global brand manager, Susan Avarde of Citigroup, and other corporate powers, the mark will arguably remain one of the most seen and recognized logos in the world. Recent events as reported in the New York Times by Eric Dash suggest that the changing culture of Citigroup may undergo additional identity changes. The company is organized into three major business groups— Global Consumer, Corporate and Investment Banking, and Global Wealth Management— in addition to one stand-alone business, Citigroup Alternative Investments. Respectively, there are 14 distinct business entities, some of which utilize the new design while others continue to employ the use of the original Travelers umbrella as part their individual identities.

The original red umbrella from the Travelers logo remains in use primarily by Citigroup and several of its brands; keeping the umbrella apparently was Reed's solution before the development of the new logo during the transition period. The red umbrella, tucked to the right of the logotype that reads Citigroup, is disconnected and a poor fit for the new sans serif font of the logotype. As a graphic element, it worked much better with the serif font of the original Travelers logo for which it was designed. I question its value and continued use in terms of the visual equity of the brand.

The Travelers division of Citigroup has since spun off as its own separate company and through a recent merger with the St. Paul Company may soon be a substantial competitor of Citigroup. Their new logo employs the use of an oval red swooshlike shape that surrounds the word Travelers and has more in common with the ABC Sports logo than any connection to its original incarnation. Having relinquished all attachments to Citigroup in terms of identity except for the use of red, the new logo is the result of a company-wide logo contest. The

Figure 1-13.
Final Citi logo.
(See also Figure (v) in the color insert.)

current graphic design is more in line with a sports event than a direct connection to the corporate culture and the ideals of the company that it represents.

Visual Relevancy Amidst the Noise

The Citi identity redesign successfully embodied the corporate cultural and consumer brand expectations of the company. The company continues to evolve and change, as do many other corporations during these challenging times of Enron scandals and aggressive mergers and acquisitions. For designers, these mergers will continue to provide opportunities for successful visual communications with challenging and important roles in the development and application of products and brands for the consumer. Design is no longer this "other thing over there" as stated succinctly by Scher when discussing her earlier experiences

Figure 1-14.
Applied Citi logo design | Credit card.

as a designer learning to negotiate the corporate landscape. Challenges abound, but Scher and her Pentagram team have remained positive and enthusiastic about the process—and the results. This speaks to a professionalism that clearly avoided the political minefields that were part of the merging of two companies into one. Pentagram remains the keeper of the mark, working closely with both Lippincott Mercer and Fallon, in the usage and implementation of the logo in its many media applications.

The high points of the Citigroup project and the creative process that developed one smart and effective design are in the genius and

Figure 1-15.
Applied Citi logo
design | ATM card.

Figure 1-16.
Applied Citi logo
design | ATM machines.
(See also Figure (vii) in
the color insert.)

strategy of the presentation by Michael Bierut and of course in the visual relevance and beauty of the design itself. I have witnessed the visual relevancy of the new mark and its power to connect to consumers through the successful application for consumers as far away as Shanghai, China. Given the compositional strength and the simple beauty of the mark, it is clear to me that its historical relevancy as a design solution will continue to grow and be discussed despite the unpredictable and ever changing nature of the corporate giant it represents.

Chapter 1 Summary and Exercises: *Navigating Corporate Divergence*

Paula Scher and the Pentagram creative team she headed understood fully the scope and implications of working with the newly merged corporate giants Citicorp and the Travelers Group. This merger was tagged by Forbes Global as the world's largest company and the most profitable financial services firm at the time. Citigroup, the first U.S. company to combine banking with insurance services since the Great Depression needed a corporate identifying mark that would be just as visually relevant in the landscape of marks and logo-types as the company and corporate culture it would come to represent.

1. A Designer's Keys to Success

Arriving at what proved to be the critical first step in the approval process in the development of the new logo was the genius behind the presentation of ideas spearheaded by Scher and her Pentagram partner Michael Bierut; they grouped several of the world's most recognizable logos on the same board with the newly designed "Citi" logo. This essentially created the culture (for the presentation) of the bank of the future, which included the specific profile of the bank's customer—an eleven-year-old girl

whose notion of banking services differs greatly from that of today's consumers. Her priorities are ease of access as well as the ability to do her banking from anywhere in the world.

2. Visual Relevancy and the Art of the Successful Presentation

The presentation effectively dramatized the new design's relevance alongside logos that have been a part of our visual culture for decades. This creative strategy emphasized the notion of "forward thinking" on the part of the client, which led to the approval of one of the most widely recognized and memorable corporate logos designed in the last twenty years.

3. Corporate Decision Makers, Language, Human Nature, and "The Red Dress"

Knowing the key players at the table during the review and decision-making process was critical for Scher and the Pentagram creative team—and to all designers in professional practice. This information, a keen observation of human nature, and the willingness to embrace change are other essentials in the process. Or as Scher put it, you need to persuade the client "which has been wearing the same blue dress to the dance to suddenly change and decide to wear a bright red one instead. . . ."

The designer needs to decode the process through the use of language in describing creative options to the client and to understand the changes that the clients are willing to make given their corporate culture and accepting them—blue dress, with a red belt.

Design Exercise

Review several existing corporate logos that have been part of our visual culture for more than ten years and that you feel require

*revitalization or a complete redesign.
Thoroughly research the company and
acquire knowledge of the corporate structure
and culture of the organization.*

- *Update the existing logo through new
  approaches to its application, i.e., General
  Electric, FedEx, UPS*
- *Design a new logo for the company*
- *Apply the new or updated logo to all
  appropriate items specific to the company's
  consumer products and services for
  creative review*

Professional Practice

*Evaluate existing approaches to client
presentations assessing the effectiveness and
the approval process of the concepts.*

- *Incorporate presentation strategies that
  will dramatize your approach while clearly
  validating the effectiveness of your ideas*
- *Know each decision maker in the approval
  process before the presentation*
- *Incorporate design language that is
  human—and easily understood by the client
  without diminishing the power of your ideas*

Intuitive Design and Culture: "East Meets West"

Alan Chan Design—Hong Kong

New Coca Cola Logotype—China and Mr. Chan Tea and Restaurants—Hong Kong

Figure 2-1.
Alan Chan.

and provocative use of visual iconography both familiar and new as a vehicle to move a message forward. This approach, however, is not without the complexities attached to the use of imagery that is too familiar, or too limiting, from the viewer's perspective. Alan Chan effectively utilizes culture in what he describes as his "East Meets West" approach to visual communications, and this cultural component is central to his creative process. Chan's design reach is as broad as the neon lights of Hong Kong are bright.

Chan has created tea rooms in Japan; restaurant, boutique, and nightclub interiors on "The Bund" in Shanghai (including the logo designs for each venue); Olympic posters for Beijing 2008; his own merchandising "Alan Chan Creations" (the label he launched in 1990); and the "Chan Tea" beverages. He also appeared in the nationally televised commercials promoting the drink in the featured role of the "Tea Master" for the popular Kirin beverage company based in Japan.

The Coca-Cola Company chose Chan to create a version of their world famous logo based on traditional Chinese characters for use on the Chinese mainland. The new logo, whose letterform construction would successfully merge the nuance and detail of traditional Chinese characters—and yet retain the visual branding of the legendary Coca-Cola script—was launched in 2003 among the fanfare usually associated with pop stars. This particular example of the talent and resources of more than thirty years in the graphic design and advertising business confirmed Chan as a designer of international merit.

The difficulties associated with creating the new logo for Coca-Cola that are expounded on in this chapter also highlight Chan's creative process and affirm his place as a leader and visual communicator in the complex and growing field of graphic design.

Introduction

How significant is the impact of culturally derived messages on the effectiveness of marketing and visual communications targeted to the general consumer? Does the use of culture as messenger have the power to move diverse markets without alienating large segments of the population unfamiliar with its modus operandi? The quantitative information needed to truly answer these questions will not be provided here. However, what will be reviewed is how one's cultural background can increase the size of the creative canvas, thus allowing for thoughtful

Figure 2-2.
Hong Kong western market.

Overview

Chan was an only child of parents who migrated in 1950 from mainland China to the former British colony of Hong Kong. His father sold fruit on the streets, eventually saving up enough money to open a shop of his own—modest origins for Hong Kong's most successful designer. He never received formal training in art except for one evening course; he is a self-taught graphic designer. Although his focus in high school was mathematics and science, he realized during his teens that he was drawn to the visual arts. He attributes his desire to design and create things with his own hands to watching his father, who made most of the items they needed for their daily existence. His father made furnishings such as tables, chairs, and

Figure 2-3.
Ying-Kow Road, Hong Kong.

Figure 2-4.
Central Hong Kong.

Figure 2-5.
Actors Maggie Chan and
Tony Lueing from the
Hong Kong feature "In
The Mood For Love."

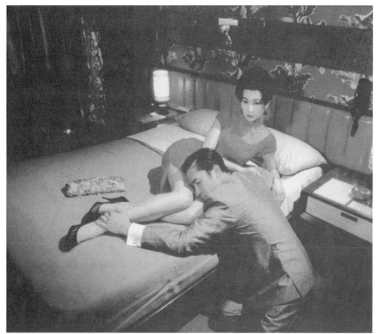

benches from the emptied wooden fruit crates. Chan remembers with pride and great detail the energy and creativity that his father put into his own creations—often deconstructing the fruit crates to make tables and cabinets—then repainting and adorning them with working Art Deco fixtures for use in the family's home. Chan sees this as a

direct link to his early interest in design. His own aspirations had been planted by the best kind of cultivator and in the most fertile of soils, that of a father to his son.

Hong Kong, or HK, as it is referred to by some of its citizenry is a center of world trade, banking, and manufacturing, and became the natural meeting point of East and West on many cultural and business fronts. The city remains a thriving urban hub full of contrasts, from extreme wealth to abject poverty—all within a storefront's distance. The density of the city is immense, and it is a challenge to navigate the teeming sidewalks. This mass of humanity is equaled only by the visual assault of neon and billboard advertisements and the promotional designs that permeate the urban landscape.

Having quietly transitioned from Great Britain's rule (dating back to 1841) to the People's Republic of China in July 1997, Hong Kong functions under the title of "Special Administrative Region." It remains unique in its approach to business. The transition agreement was signed twenty years ago by both China and the United Kingdom. The "one country two systems" formula, which is the current structure of Hong Kong, will be in place for another forty years. At that time it will become a full participant in China's national economy.

The western influence in HK is as conspicuous as its crowded streets—from the sophisticated clothing styles with designer labels like Prada and Gucci, to contemporary film, music, art, and architecture. The city's mix of contrasting styles, neon lights, street signs, and lamp posts that also serve as clothes' lines form the environment in which Chan's fusion of Oriental and Western sensibilities came to fruition.

A prime example of this mix is the film "*Hua Yang De Nian Hua*" released in the United States under the title "In The Mood for Love" in 2002. The film is the product of Hong Kong's outstanding filmmaker Wong Kar-Wai. The Hong Kong-based director weaves a tapestry of deceit, love, and longing in the hot-pot of 1960s' Hong Kong. Infused with the Cantonese and Shanghainese dialects typical of that time, Wong Kar-Wai's deliberate use of sophisticated fashion in the slim-cut men's suits give the lead actor Tony Leung the cool understated screen presence of a jazz aficionado. The contrast between and metaphor of the male–female relationship, as expressed in the understated but powerful Cheongsam dresses worn by co-star Maggie Cheung, give the film its captivating visual presence. It is not surprising that the director was a graphic design student at Hong Kong's Polytechnic University. This power of contrasting elements is at the very essence of successful design. The design principal of ABA form (repetition and contrast) is clearly and successfully at work here.

The film is shot in a uniquely visual way that is unequaled, and the soundtrack uses Pingtan, Cantonese, Beijing, and Zhejiang operas alongside songs such as Nat King Cole's silky smooth rendition of "*Aquellos*

Figure 2-6.
Alan Chan Design Co. logo.

ALAN CHAN DESIGN CO.

Ojos Verdes"; songs from mainland China like "*Shuang Shuang Yan*," performed by Deng Bai Ying; and the "*Bengawan Solo*" performed by the great Rebecca Pan. This mix reflects the "oriental passion and western harmony" that permeates Hong Kong popular culture, and which Chan continues to inculcate in his designs.

The western influence is noticeable in advertising and graphic design in Hong Kong. This is not to suggest that the Oriental mindset is invisible—rather it allows for a more apposite review of the playing field that Chan has successfully navigated as a visual artist and communicator. The projects in this chapter will be reviewed contextually and will deal specifically with graphic design, advertising, and merchandising, from the power of the "ethnic visual nuance" to Chan's approach to visual communications and the creative process.

"A Thousand Beautiful Things"

I had the good fortune of meeting Chan during a visit to China in the spring of 2002. That was the first time I became acquainted with the man who is arguably the most successful graphic designer Asia has given the world. The occasion was a birthday

Figure 2-7.
The Hong Kong Museum of Modern Art's "The Art of Living" Alan Chan exhibition 2003.

dinner for Chan. The dinner took place in a restaurant located in the heart of Shanghai's hippest and newest nightclub and dining district, Xintiandi. As the evening unfolded Chan made it a point to speak with me, discussing his background and current projects in spite and of the occasion and presence of his friends and colleagues. I was amazed at the sheer depth of his design work. While he had been featured in several design publications, somehow I had missed them and the enlightenment meeting him

Figure 2-8.
The Hong Kong Museum of Modern Art's "The Art of Living" Alan Chan exhibition 2003.

now provided. Here was a self-taught designer with more than three decades of experience in all forms of visual communications—a designer who B. Martin Pederson, the publisher of *Graphis* magazine, proclaimed headed one of the "ten best design firms in the world."

My first impressions of him were those of a refined and gracious man, articulate and confident with a polite modesty. He accepted my gratitude and heartfelt compliments on such a prolific career. After discussing my background and the nature of my visit to China he gave me a book that highlighted his illustrious design career. The book was part of a design series featuring the likes of Woody Pirtle and other standout creative thinkers of contemporary graphic design. Chan agreed that design is essential for the continued success and progress of our society. Creative people responsible for the best of it know this and infuse their work with this principle. This reality is borderless. Chan's portfolio of design work covers everything from corporate identity, posters, and package and interior design—to his own licensed merchandise created and sold in museum stores and marketed exclusively in his Hong Kong boutiques. His posters and watch designs are included in the San Francisco Museum of Modern Art's Permanent Collection of Architecture and Design. The list of awards is long, international, and impressive, more than five hundred in all, from organizations such as The New York Art Director's Club, Communication Arts, The Tokyo Type Directors Club, HK 4A's, and others.

Entering the world of Chan's creative accomplishments the selection of the song "A Thousand Beautiful Things" by Annie Lennox is an appropriate soundtrack for the occasion. His body of work is beautiful and thoughtfully conceived. A year after

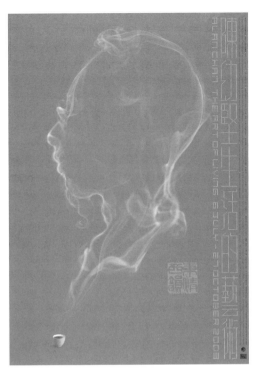

Figure 2-9.
"The Art of Living" Alan Chan promotional poster 2003. (See also Figure (viii) in the color insert.)

meeting Alan, he agreed to be interviewed for this book and became an active participant in design education and its validation as an integral part of society. The interview took place in his office overlooking the busy Hong Kong Harbor. The Chan Design firm has been in business for more than twenty-three years and conducts business throughout Hong Kong, mainland China, and Japan extensively. The company's impressive client list includes international giants like Coca-Cola and Walt Disney. The following examples focus on his creative process in arriving at solutions that both satisfy the market and meet with approval by his clients and consumers, yet are infused with the rich flavors of his "East meets West" approach to design.

Good Taste, Be Happy

The ability of words to compel consumers to participate or act when applied to advertising and communication campaigns is one of the most effective ways of stimulating

Figure 2-10.
New Coca-Cola label
graphic launch. (See also
Figure (ix) in the color
insert.)

The world is familiar with the legendary taste and look of the Coca-Cola brand with its Spencerian English styled script, glowing white with the familiar red background and flowing white ribbons or waves. However, in China the beverage routinely consumed in copious amounts is tea, not cola. Tea, which is culturally tied to all of Asia, is generally served warm with just about every meal. Taking on tea as the main competitor in the drink market—even for beverage king Coca-Cola—would be an immense goal given the reality of tea's position and its revered place as part of the day-to-day cultural ritual.

Tea will not go the way of the dinosaur—and just as chairman Mao remains an icon in the minds of the people, "Cultural Revolution" notwithstanding—tea is embraced and remains a familiar part of the fabric of Chinese culture and will continue to be an integral part of the daily lives of the people who drink it. "Its place in the culture is tied to art, music, poetry, calligraphy, and the enjoyment of life."

With this reality, Coca-Cola executives and marketers in China, and Chan himself, the company was quite content to share the table with the peoples' number one drink—tea. When the Chinese do reach for an ice-cold carbonated refreshment, that choice should be Coca-Cola. Because of this, they will continue to be successful in terms of market share and brand loyalty in China.

The new graphic was launched in February 2003—the Year of the Ram. The unveiling took place at the Shen-Mei bottling plant located in Shanghai. In attendance at that time were company president Paul Etchells, vice president and marketing director John Cheung, vice president (China Operations) Atul Singh, and general manager of the Shen-Mei Beverages and Food Co., Ltd, Mike Beale. Appropriately Chan joined the group as the Chinese media covered the event with all the pomp and circumstance befitting

consumer markets regardless of the specific language. However, when translated from one language to another, the meaning may not be the same. Even when it is close, it often makes for strange, or even awkward syntax. Its true definition essentially becoming, dare I say it—"Lost in Translation."

Coca-Cola's Chinese launch of its modernized and culturally connected logo executed in bold Chinese characters, the literal translation informs the consumer of the state of mind that the great taste of cola brings with each refreshing drink. The direct translation of the words "Coca-Cola" in the traditional (common language) of China means "Good Taste, Be Happy" as it appears on the product with the word "Soda/Gas Water" appearing below and to the right of the new logo and set in smaller characters. The latter word serves the very practical purpose of informing the consumer as to what exactly "Tastes Good."

Figure 2-11.
Coca-Cola new Chinese graphic detail.
(See also Figure (x) in the color insert.)

an opening night performance. There was a 30-meter mosaic of the new logo, which was draped on the wall of the bottling plant in Shanghai and was accompanied by tri-colored fireworks using the graphic's base colors.

The planning and strategy for the launch was well orchestrated and part of a year-long marketing blitz called "The Year of Coke" in China. The public was becoming more and more comfortable with the idea of drinking

carbonated sweet drinks on ice with everyday meals and on special occasions—and Coca-Cola remains the best-known brand in China and the world. The Coca-Cola Company markets four of the world's top five soft drink brands, which include Diet Coke, Fanta, and Sprite. The company has spared little in the way of expense to continue moving the brand forward. Coca-Cola and its partners have invested over US$1.1 billion in China since the late 1970s. Locally, the wholly owned Coca-Cola (China) Beverages Ltd., and its bottling companies employ 20,000 Chinese citizens. The company is a major player in the continually growing Chinese economy.

The Art and Transfusion of Chinese Calligraphic Characters

In modernizing and developing the new logo, Chan sought to build on the tradition of the brand and the unique qualities of the Spencerian English script that has become the global symbol of Coca-Cola. This marks the first time since 1979 that the Chinese script is being updated on the brand. Building on the fluidity of the original logo, Chan utilized the curves and subtle stroke weights of each character while adhering to the specific shapes of traditional Chinese characters that form the graphic image of the brand. The stroke weights are demonstrably heavier than previous

Figure 2-12.
Coca-Cola launch celebration.

incarnations and generally signal more motion as well, tapering off to suggest the serif-like form of the original stylized script.

This fluidity accentuates the boldness of the new logo. Furthermore the new graphic is emboldened by the use of additional colors—these lines, which appear on the contour in dark red, create a sense of depth and underscore the aforementioned movement Chan has imbued in the new logo—a font

Figure 2-14.
Alan Chan creations. (See also Figure (xi) in the color insert.)

which he painstakingly designed and that clearly pays homage to the original. Also the use of the famous script embellishments or waves used in multiple locations over key characters in relationship to the overall graphic reinforces the brand's connection to the original script. These subtle changes would be meaningless without the accurate use of the modernized Chinese characters designed specifically for the logo by Chan. The preeminent visual communicator Hans Schleger stated decades ago "The best marks are seen and recognized—not translated. Their impact deep, becoming permanently absorbed by the receiver."

The classic Coca-Cola trademark is just such a mark. Recognized as a legitimate classic in the popular American culture alongside hot dogs, apple pie, and baseball, and as such, it is timeless. The use of Spencerian English script could have relegated it to a certain period of time, but the strength of the brand and its longevity make it an exception to this rule.

The existing design and brand strengths previously noted should not suggest that Chan's task was made easier given the visual equity of Coca-Cola. There was the challenge

Figure 2-15.
Alan Chan creations for San Francisco Museum of Modern Art. (See also Figure (xii) in the color insert.)

of designing a totally new Chinese character that, given the appropriate visual properties such as stroke weight both horizontally and vertically, predisposed the new font to the modernization that Chan was after, that of transforming the logo to a more "timely" set of letterforms while retaining the fine distinction of stylization informed by the Spencerian English script of the original mark. The new mark achieved these goals successfully without appearing forced, or as design element afterthoughts that ultimately reduce the effectiveness of any logo. Added to these strengths is the dominance of the color red, historically a pigment of good fortune throughout most of Asia. In addition, the abstract resemblance of the new calligraphic logo design to that of the stylized dragons seen throughout China provides another cultural icon associated with strength and prosperity.

These coincidental visual relationships may seem small to the casual observer; however,

in China where customs are treasured and endure for centuries—from the wisdom of Confucius to the poetic verses of the great Du Fu—these are the advantages that have assisted Coca-Cola in its ascension as the top brand in China.

"Chan—The Master of Tea"

In Chinese philosophy, the universe is balanced by five essential elements—earth, water, wood, metal, and fire—all harmoniously dependent on each other. They are perfectly balanced and linked to each other's collective existence. It is this kind of profound balance that Chan seeks to achieve in both his experiences in life and in the business of visual communications. In 1990 he launched "Chan Creations," essentially a brand extension of the Alan Chan Design Company. This marked the designer's first venture into merchandising and product design. The oriental scroll, which Alan designed as a promotional item for the firm, is over 100 feet in length and highlights his visual arts' achievements spanning three decades. The beauty and simplicity of the scroll design as a vehicle for self-promotion is a testament to Chan's creative process, which consistently reinforces the cultural positioning of his work. When viewed along with his other products and promotions, it is a resounding validation of his understanding of branding and image. He carefully promotes himself, always reinforcing his approach to design while keeping his image and persona in the mindset of potential clients, marketers, and consumers. Just as he effectively utilizes the Buddha's hands, his image has become an icon in its own right. The profile of the bespectacled (in silhouette) "Master of Tea" is the primary graphic on a range of communications devices Chan has used to promote himself, and by extension his businesses.

Figure 2-16.
Chan's tea toom.

His extensive line of Alan Chan Creations include Chinese tea, tableware, T-shirts, watches, stationery, and other items. In addition to the creation of these unique items, Alan also opened two shops in Hong Kong. This authentic line of merchandising is also available at fine stores throughout Asia, Europe, and the United States. In venturing into merchandising, management, and sales, Alan has successfully reinvented himself. The design business fuels the merchandising, which in turn reinforces the Alan Chan Brand in the mainstream. Design has become his means to an array of successful endings.

Chan's true understanding of the fusion of Hong Kong's two cultures manifests itself readily in his products, which underscored these experiences. This would continue to be his creative direction—creating items with a nostalgic nod to Hong Kong and China's rich cultural past. Each product is a celebration of Chinese culture, right down to the Chairman Mao coasters and Calendar Girl biscuits sold in his stores. The attention to detail is evident

on items like business card holders and boxed sets of tea. At the heart of his creations and residing in Chan's soul is his understanding and appreciation—he embraces tea's revered place in Chinese culture. As a consequence, many of his products and designs celebrate this notion. There is an old Chinese saying that identifies the daily necessities in life—rice, salt, oil, fuel, vinegar, soy sauce, and tea. The tea-drinking custom has been a part of the culture and has been ritualized and engendered in China for more than a thousand years.

In 1993 Alan Chan created his first tea room. Conceptually, the birth of "Chan" as an idyllic master of tea and patron of the tea rooms that bore his name came out of his understanding of the history and social relevance of tea rooms. They were known as a meeting place for intellectual discussions and centered on the advancement of art, poetry, and social issues in China dating back to antiquity. Appropriately the first tea room would be located in Hong Kong. This gave Chan the opportunity to fully engage

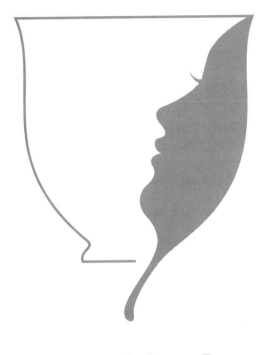

茶佳人

T E A S I T E

an Alan Chan Concept

Figure 2-17.
Tea Site logo.

his design sensibilities and focus on identity development, interior design, product development, and packaging design—right down to the design of place settings and tableware.

The logo developed for the first "Chan" tea room graphically depicts the Buddha's hand delicately holding one tealeaf. The graphic is executed beautifully; a line drawing that evokes the grace and beauty of hands seen throughout Chinese culture on a multitude of paintings and sculptures depicting the Buddha.

Alan Chan's Tea Rooms would serve another purpose as well. His fondness for the old days of Hong Kong's thriving "Dai Pai Dong," (the sidewalk or roadside stalls) which had vanished from the landscape. These had been a thriving part of Hong Kong's historical dining experience and would inform the creation of his original "Chan's" Tea Room. This ideal manifested itself both in the physical surroundings of the establishment as well as in the wonderful delicacies Chan's Tea Room offered to its customers—thoroughly evoking the very best of Hong Kong's past. The tagline created for the advertising campaign in the initial launch of the tea room nails the objectives of the experience succinctly: "It's everything you'd like to remember—and nothing you've ever seen."

The atmosphere he created would once again be fused by the colliding flavors of the eastern-and western-styled dishes offered and showcased on the menu.

You can draw a straight line from the experience and success of Chan's first tea room, which opened over a decade ago in Hong Kong, to the demand to replicate his successes in Japan, creating what he termed an "Oasis for Peace of Mind." Chan's approach to the design of these spaces—beginning with the Lantern Tea Room, Alan Chan Tea Room, and the Cha Kai Yan Tea Site in Japan—was all encompassing—from napkins to signage and overall identity. Opportunities in design that give the designer total control over the direction of all elements on such high profile jobs is unusual. This is also a testament to Chan's success and respect as a designer of the highest merit. His reputation has placed him at the top of the industry and in countries like Japan, which hold particularly high standards for design. It is generally considered an environment that would inculcate design excellence in the mind-set of its people.

As a brand, "Alan Chan" would be extended and summoned out of the tea room and onto

Figure 2-18.
Exterior of Tea Site
project.

Figure 2-19.
Tea Site interior
design.

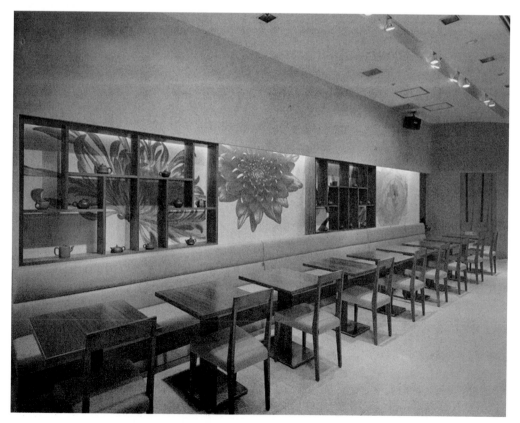

Figure 2-20.
Lantin tea house menu.
(See also Figure (xiii) in the color insert.)

the countertops of the grocers throughout
Japan and Hong Kong—introduced to
consumers in cans and bottles designed and
packaged by the "Master of Tea"—Chan
himself. This new beverage was launched and
marketed exclusively by Kirin Beverage
Company, Japan. Chan was approached
directly by Japan's third largest advertising
agency ADK Asatsu-DK Inc., with the idea of
pitching the Kirin Beverage Company to
develop a new tea drink based on "Mr.
Chan"—the Master of Tea. Alan agreed and
Kirin enthusiastically accepted—providing
him with an opportunity to develop the
brand, design the packaging, and in what
amounts to a slam-dunk business deal, sell
the licensing use of his image "Mr. Chan" to
the company as well. That in and of itself
puts Chan in a category by himself.

The success of Chan's original blend of
tea for Kirin led to national advertising
campaigns in print and television featuring
himself as the Master of Tea. The spots were
shot on location in the ancient capital of Hue

in Vietnam. The creative direction and
production of these spots were developed by
the advertising agency ADK of Japan.

Identity Design

The list of projects Chan has completed in
Hong Kong and Japan is significant. As the
business climate continues to change, it
would seem that future opportunities for the
design company would come from mainland
China, as their economy continues to show
unparalleled growth. Recently completed
projects include "Three on the Bund"
located on Shanghai's legendary Zhongshan
Road known by travelers simply as the
Bund. Alan Chan not only designed the
identity for the building, but also developed
identity programs and, in some cases,

Figure 2-21.
Lantin tea house
printed items.
(See also Figure (xiv) in
the color insert.)

Figure 2-22.
Lantin tea house
signage.

is an extremely effective mark—clearly projecting an unusual image of a city skyline while maintaining the needed legibility as logo and signage vehicle for the establishment.

The building is one of many that date back to the early 1900s, when the United Kingdom and Japan ruled certain areas of Shanghai exclusively. Many of these buildings were financial institutions. This particular building is now the home of fine dining, shopping, and health spas, which bears the distinct signature of Alan Chan's design aesthetic. All signs point to the mainland for the continued business prosperity of Alan Chan Design.

When given an exclusive high-end customer, as in the case of Seibu, the chain of fine department stores out of Japan, Chan's sense of design and the refined cultural fusion of east meets west was readily wrapped in the delicate visual nuance and simplicity that is appropriate for this prominent client from Japan. The store's location in Hong Kong's exclusive Pacific

interiors for many of the tenants residing there. Of particular interest and from a design perspective the "New Heights" logo

Figure 2-23.
Lantin tea house
interior. (See also
Figure (xvi) in the
color insert.)

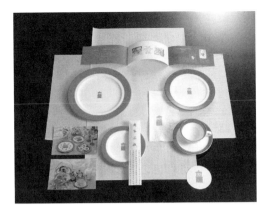

Figure 2-24.
Lantin tea house flatwear design.
(See also Figure (xv) in the color insert.)

Figure 2-25.
Mr Chan tea products (Kirin Beverage Co., Japan).
(See also Figure (xvii) in the color insert.)

Figure 2-26.
Three on the Bund logo design for New Heights.

vitality, happiness, and affluence. The stroke weights of the fish are brush-like and indicative of traditional Asian drawings and paintings. The logotype for Seibu was created by Japan's renowned graphic designer Ikko Tanaka. The swimming carp are reminiscent of ancient Chinese carvings in jade and wood. The use of these elements on the various packaging designs created by Chan effectively creates the upscale look and feel appropriate for the client.

Intuitive Design

These visual relationships to culture and lifestyle are not coincidental. They are a constant reminder of Chan's creative process, which successfully utilizes diverse cultures in the sublime fusion of his east

Figure 2-27.
Alan Chan researching artifacts for Seibu logo design.

Palace demanded that the oriental passion that is unique to Hong Kong be a key element to the mark and its application on packaging and products as well. The logo is the epitome of beauty, culture, and simplicity. The mark is based on two interlocked carp, swimming in a graceful circle forming the letter "S" of Seibu. The use of this bilateral symmetry is the symbolism of the fish itself, representing abundance as does the word "fish" in Chinese and consequently implies the notion of prosperity. The motion, perpetually implied by the design also reflects

Figure 2-28.
Seibu dimensional logo.
(See also Figure (xviii)
in the color insert.)

Figure 2-28.
Seibu dimensional logo.
(See also Figure (xviii)
in the color insert.)

Figure 2-30.
Seibu in-store signage.
(See also Figure (xx) in the color insert.)

Figure 2-29.
Seibu packaging.
(See also Figure (xix) in
the color insert.)

meets west approach to design. The skillful use of elements that have historical relevance along with the cultural content reflective of the period and market are shrewdly applied and are a central component to his business acumen as a creative thinker, problem solver, and visual communicator. With the Seibu logo, and its application to the merchandise and shops within the chain, Chan's designs are informed by his knowledge of Chinese, Japanese, and Western cultures. He has built his practice on this ability which is not only the result of years of professional practice, but is based on his intuitive ability to apply his approach to visual communications without the appearance of being dated, condescending, or out of step with the times. An in-depth knowledge of many cultures is required to achieve this. This approach is not without risks inherent in visuals that reflect and utilize specific cultural iconography and ethnic depictions as a means of conveying a point of view. His attention to detail harkens back to his studious observations of his father. Chan was attentive to what pleased the shoppers—both Westerners and the local mix of Canton and Shanghainese sensibilities—who were eating and buying the marvelous merchandise in the local *dai pai dong* of his youth.

Chan continues to travel this road successfully, avoiding stereotypical solutions to design problems and the uniformity of visual mediocrity. He has set a high standard

Figure 2-31.
Seibu credit cards.

for those who aspire to follow his lead—the lead of a master of design.

Chapter 2 Summary and Exercises: Culture as Visual Messenger

Alan Chan's deliberate use of cultural iconography and imagery in the creation of successful design solutions is at the epicenter of his decades' long success. His creative range encompasses corporate and product design, advertising and interior design, and fine arts. His visual perspective and creative process are the result of the eastern and western influences of Hong Kong's cultural melting pot— Cantonese, Shanghainese, and British.

1. Designer's Keys to Success

Alan Chan's ability to create contemporary design aesthetics with visual imagery and iconography from one of the oldest surviving cultures (mainland China and Hong Kong) while decidedly driven by sound strategy, is a reflection of his own personal history and

rich cultural background. Understanding the important connections that consumers have in the thriving marketplace of Hong Kong, and infusing it with local and traditional customs—Alan Chan blends Oriental and western sensibilities that move his ideas and the products they represent forward.

2. The Tea Room Master

Alan Chan's understanding of the history and social relevance of tea rooms as the central meeting place for intellectual, social, and artistic issues is at the foundation of his success as one of the most sought after designers of this popular retail venue. In the development of the many tea rooms he has created (dating back to 1993), while visually and thematically unique, each environment pays great attention to the details of a practice that has been ritualized for more than a thousand years. He essentially creates an atmosphere where the consumer is the most comfortable with and on some level, identifies with, the perfect environment for the introduction of new products to a captured audience.

3. Placing Products in the Buddha's Hands

Alan Chan utilizes powerful cultural icons and imagery that date back to antiquity to promote products. Information is the central thread running through his creative process. The connection he makes to the Dai Pai Dong *from Hong Kong's past evokes the best of that rich history and provides customers with the fused atmosphere of eastern-and western-styled products. The success of this approach would not be possible without his sincere reverence and understanding of the culture he delicately exploits in the promotion and marketing of products and services to the consumer. The manifestation of the above is underscored by his outstanding conceptual and executional ability.*

Design Exercise

Select or create a new product for the general consumer market whereby the use of specific cultural iconography is used as a visual strategy in marketing the product to the consumer.

- *Develop the essence of the brand: Use urban culture, i.e., hip-hop or street personae, indie, etc.*
- *Select the most appropriate venue for the promotion of the product, i.e., print advertising, billboards, Internet, etc.*
- *Develop a promotional campaign.*

Professional Practice

Evaluate your ideas about popular and historic culture and the opportunities that may exist in your practice as a strategic approach in meeting the goals of your clients.

- *Reach out to potential clients where culture plays a central role in their mission, product, and/or service to their target audience.*
- *Where appropriate, incorporate strategies using culture to dramatize and validate concepts in the marketing of products and services for clients.*
- *Thoroughly research all issues pertaining to the specific creative direction and use of all cultural and historic iconography being considered in the marketing of the products or service; assure a high level of integrity and respect so that the client and their constituents' ideals are never compromised.*

Graphics in Motion:
Handmade Design

Kyle Cooper—Principal/Creative
Director Prologue Films

Main Title Sequence Design—David
Fincher's 1995 Feature Film "Seven"

Main Title Sequence Design—Sam
Raimi 2002 and 2004 Feature
Spiderman I & II

Figure 3-1.
Kyle Cooper.

For Kyle Cooper, the creative mind behind the opening title sequence design of the Fincher thriller, it signaled the professional breakthrough that would propel him forward in motion graphics and specifically film title sequence design—his name would be associated with ground-breaking excellence. Not since the successful and innovative career of Saul Bass, who blazed the film title sequence design trail with feature films like "The Man with The Golden Arm" and "Vertigo," had one name become synonymous with the visual power to compel an audience and the feature film it introduced forward. Bass's title design work on feature films like the previously noted Hitchcock psychological thriller "Vertigo"—which told the compelling story of one man's obsession with his memory of a woman, coupled with the protagonist's abject fear of heights—continued to introduce the theatregoing public to films of the popular culture in a visually stunning and thought-provoking way from the mid-1950s to the 1990s.

The complex and slow-moving story line, which framed the action on screen in "Vertigo" still resonates as much today, as it did when it opened in theatres in 1958. In developing the title sequence for the breakthrough film "The Man with The Golden Arm" for the director Otto Preminger in 1955, Bass essentially established the credibility and importance of design in titles for feature films. He would also be instrumental in the promotion of the films—creating poster designs used in national advertisements for the movie's premier. In reviewing the design for the poster, Bass's design influence is obvious as his graphic depiction of the constricted human arm of the addicted musician, performed surprisingly well by a young Frank Sinatra, is reminiscent of the great poster designs of the 1930s and 1940s that emerged throughout Europe—particularly

Introduction

In 1995 the film industry and the theatregoing public were given a gift—an unsettling and arresting visual title sequence design introducing one of the most disturbing films since "Silence of The Lambs" and "Rosemary's Baby." The Film "Seven" produced in 1995 was directed by David Fincher and featured the talented actors Brad Pitt and Morgan Freeman working together for the first time. The film opened with a vivid and disturbing imagery imbedded in the narrative of two detectives as reluctant partners whose lives run in parallel in pursuit of a sadistically brilliant serial killer. The title credit design was a gripping visual dialogue between the viewer, the beginning of the story, and the bizarre mind of the killer, played by Kevin Spacey.

from the Swiss designers. These designers, using type and image, were able to successfully bridge the divide between abstract and representational form on the page. Their visual strength and composition are unsurpassed.

Saul Bass was the first to truly integrate graphics, provocative use of typography, and image in the design of film titles for major motion pictures. The films "North by Northwest" (for Hitchcock) and his last film title design project "Casino" for Martin Scorsese, are classic examples of his approach. Scorsese's films, including "Goodfellas" and the critically acclaimed "Casino," confirmed that Bass had not lost a step in returning to film title design after a long hiatus during which he focused primarily on graphic design and advertising along with producing his own films. He has developed corporate identity programs for corporate giants like Bell Telephone (before the break-up), United Airlines, and AT&T. These logos are familiar to the general consumer and have become part of the American iconic corporate design landscape.

Director Martin Scorsese confirmed the importance and power of title sequence design in film as practiced by Bass. "A mini film within a film—his [Saul Bass] graphic composition in movement function as prologue for the movie. His approach created tone and provided the mood—foreshadowing the action on the screen."

In his statement Scorsese got it exactly right—the opening film title sequence sets the table for the viewer. When designers are successful in this medium the audience is thoroughly "primed" for the visual experience that unfolds before them. The designer effectively stimulates the many sensory preceptors resident in the audience. Clearly this is not lost on one of the great film directors of our time. Scorsese's choice of words, specifically the word prologue in describing the Bass approach to film titles is in fact the moniker of Cooper's new company "Prologue." Serendipitous? Perhaps.

As fate would have it Martin Scorsese would have an opportunity to work with the man who was on his way to becoming the "first name" in title credit sequence (TCS) design. Cooper was originally hired to design the TCS for the film "Goodfellas." During collaborative discussions with the director Cooper's inability to connect with the look and feel the director wanted for this dark photoplay about wiseguys in organized crime had delayed approval of the TCS design and the film was nearing completion. This was not due to any creative ideological chasm or any ethical shortcoming on Cooper's part. It was merely, and frustratingly from Cooper's perspective, a case of his inability to present the director with ideas that resonated and underscored Scorsese's vision for the film. During a casual conversation on the set, Scorsese was relating a story about TCS design to Cooper in an effort to facilitate the creative thought processes in the hopes of arriving at a concept for the movie. The example he highlighted happened to have been designed by the famed Saul Bass, whom the director thought was deceased (at the time Bass was alive and well). When Cooper informed Scorsese that in fact the renowned designer was still alive and working, Cooper was promptly dismissed and replaced by Bass. His attitude about this experience speaks volumes and is a testament to his wisdom and success.

When Martin made the change I accepted it as the professional decision that was his prerogative as the director of the film. Whenever I am working with a director I never lose sight of the fact that I am part of the film's crew, and as such I will always be a team player. At the end of the day I want what is best for the film as described by the director whose vision defines the narrative as it should—even if that means being replaced.

Overview

The creative development of Cooper is as compelling a story as some of the feature films he has designed for. In reviewing his work and his educational experiences and influences over the past twenty years, the relevance of postmodernism and its impact on design, as well the Bauhuas and the Swiss design that came out of the Basel school, garner more than a nod by this astute and well-studied design practitioner. These experiences clearly inform his approach to the use of type, live action/image, and form in his approach to motion graphics. Interestingly his primary design influences are not from creative thinkers in time-based mediums. He speaks articulately and unabashedly about his reverence for Paul Rand, and more

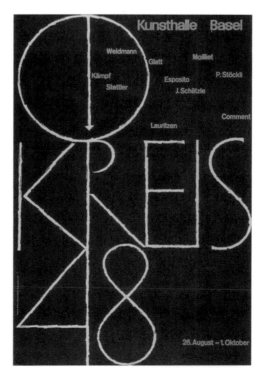

Figure 3-2.
Armin Hoffman poster design.

Figure 3-3.
IBM Cover by Paul Rand.

notably the sincere friendship that developed out of that student/mentor relationship during his years as a Yale graduate student. This relationship would continue until Rand's death. Paul Rand, noted for his ability to be a caustic critic and curmudgeon as well as his status in the field of design, holds an endearing place in the life of Cooper. These experiences solidified the importance of a practical approach to design along with the distinct creative perspectives imparted on him by his mentors, some of whom came out of the famed Basel School in Europe led by design educational luminary Armin Hoffman.

In highlighting the outstanding career of design auteur, Cooper, in this chapter we can clearly see that his ascension to prominence was no coincidence. It was borne out of the culmination of hard work, planning, and focused educational pursuits that refined the natural talent that Cooper possesses—morphing much like some of his font animations into a motion graphics career that continues to move in one direction—up.

Over the last decade Cooper has left a mark on both large and small screens—infusing live action image and graphics with a mix of classic letterform application and juxtaposed elements reflective of postmodern influence and popular culture—inevitable comparisons to his predecessor notwithstanding. His professional list of clients is impressive. It includes design and live action concepts for companies like Mercedes Automotive and Olympus Cameras; film titles and special sequence design and second unit live action concepts for the HBO Emmy Award winning dramatic series "Angels in America"; and feature films like "Mission Impossible," "Men in Black," "Donnie Brasco," both Spiderman films, and his magnum opus "Seven." He has also created design and animation for Apple iTunes Europe, MTV, and Hewlett Packard. His diverse professional odyssey effectively bridges the gap in motion graphics in what he calls his handmade approach to design—utilizing his experiences working under legendary design masters Paul Rand and Armin Hoffman—an educational experience firmly rooted in two-dimensional graphic design work and not time-based mediums.

That he would walk into the most progressive motion graphics and animation studio in the industry with little experience in time-based work, and successfully walk out blazing a trail that would garner the most notable form of acceptance—that of replication—is a testament to the talent and visual mettle of Cooper.

He is a creator of visual bridges, as he is now arguably the standard bearer for film title sequence design in the industry. He executes

Figure 3-4.

Mission Impossible title sequence.

design solutions that retain a clear connection to old school visual communication principles and perspectives while inculcating some design characteristics associated with postmodernism and popular culture. He conceives and executes ideas that are as visually provocative as they are beautiful.

The Guidance and Wisdom of a Design Master—Multiplied by Two

The recognition of Cooper came after working in the field of motion graphics for more than ten years. Prior to this Cooper completed visual arts degree programs in interior and graphic design at the University of Massachusetts and Yale University. As an undergraduate at U Mass, enrolled in the BFA program, Cooper was a scholarship student whose primary focus was the discipline of graphic design in spite of the interior design curriculum that was the program his scholarship supported. "My goal was to study graphic design, but at the time U Mass did not have that specialized focus as a stand-alone area of pre-professional study."

After the completion of that program he would go on to study design at Yale, where he worked closely with the designers Paul Rand and Armin Hoffman respectively, both of whom he says helped him define his own philosophical and practical approach to design. As noted previously his relationship with Paul Rand developed beyond the typical student-teacher mentoring experience. This would leave an indelible mark in Cooper's professional growth and understanding of design. "Paul said everything your hand can find to do, do with all your might. He believed that many times the answers to the design questions were to be found within the work itself, proclaiming that one must constantly listen to the work."

Cooper was an attentive and respectful student who harnessed the knowledge of the two design masters in Yale's distinguished graduate program. This wealth of old school knowledge led Cooper in the development of his distinctly unique use of type and image on the big screen. He deliberately fused these experiences and infused them into his own approach to design. Armin Hoffmann maintains a place in the annals of graphic design and education that put him on the level of a cult figure. His textbook *Graphic Design Manual: Principles and Practices* is a staple in many design school libraries more than forty years after its first publication in 1965. In the words of author Steven Heller "His poster designs uniquely bridge the gap between abstraction and representation . . . few other designer's have produced more vivid work that can be seen from one hundred feet—or one foot away."

Hoffmann's methodology in design education has impacted curriculums from New York to Tokyo as well as Europe. Arising from Switzerland's legendary city of Basel—a haven for design excellence—with other notable designers like Niklaus Stoecklin, Burkhard Mangold, and Wolfgang Weingart, to name a few. As if having the great Paul Rand as a mentor was not fortuitous enough. Cooper was a sponge—linking the design practicality of Rand to the Swiss simplicity of Hoffman. Hoffmann's significance as a designer and educator is summarized succinctly in this quote from Paul Rand:

Design is an all-encompassing discipline, embracing the spiritual as well as the practical problems of life. The list of subjects a good design teacher is aware of ranges from the fundamentals of drawing and painting to the study of mathematics, psychology, and philosophy and to a more than passing acquaintance with the history of art and design. Hoffmann's posters for museums and for the Basel State Theatre and his architecturally related projects are evidence of his social consciousness and his preference for the kind of work that permits artistic freedom. Nevertheless, he is keenly aware that good design means satisfying practical needs. Hoffmann stresses the higher goals of art and education not to put down the world of business but to give it real meaning.

Unarguably the first name in graphic design, Paul Rand's work is indeed legendary and needs no further review here. Having his mentorship coupled with Armin Hoffmann, former head of graphic design at the Basel School of Arts and Crafts during the years 1947 to 1987, provided Cooper with a one-two punch in visual arts education that is and remains unparalleled. Cooper's success pays homage to the wise and talented "trees of wisdom" from which he cultivated his talent during his Yale experiences. Often these experiences had divergent paths as with the super modernistic and practical approach of Rand juxtaposed to the Swiss design of Hoffmann where the seeds of postmodernism in graphic design began to take its earliest roots. Cooper's talent and success is much like the fruit that does not fall too far from the trees.

After graduation Cooper began a professional odyssey that would take him to the pinnacle of his profession while working both coasts for more than ten years as a motion graphics designer. In beginning his professional journey he essentially started at the top, if not in position certainly by association. He began working for the famous R/Greenberg studios based in New York City, which at the time was the preeminent motion graphics and animation studio in the country. Cooper applied successfully for this position without any experience in time-based design work. While his graphic design portfolio was first-rate, showing the promise of his preparation and natural ability as a visual artist, there was no indication that he would thrive and succeed working in the time-based medium of motion graphics. He effectively talked the R/Greenberg partners into hiring him. Cooper explains: "good design is just that, good design. If you understand and embrace this fully then it is possible to work in any medium—as a designer. I'm not saying that as an example I could be a practicing architect—however as a designer I am confident that I could design a building or a house. For me the process is the same."

The previous statement is a telling one in its sincerity of purpose, and it is built on the optimism and talent of a designer who relishes the opportunity to be a designer, finding "any opportunity to be creative with your hands, whilst working as hard and thoughtfully as you can." This is one of Cooper's guiding principles, being a hands-on designer and adapting to any situation and most importantly, listening to the work. He describes the time-tested practice of his mentor from Yale: "Paul had a bulletin board whereby anything and everything he came across during the course of his experiences that got his attention [visually] he would tack up on the board as it may invariably lead him to something he could build on and perhaps use to solve a design problem. These items represented a vast array of origins and topics. From architecture, history, and technology to the current events of the day. He might not have a specific purpose in mind at the time, only that whatever it was, if it found its way onto the board and it was inherently thought provoking, and as such might lead him to something more profound in forming his approach to a design problem, he put it up there." This kind of "open" thinking, as opposed to the narrow view of looking only in areas specifically related to graphic design has become one of the cornerstones of Cooper's creative process. Walking through the designer's massive Prologue Design Studio in Malibu, California, the casual observer cannot miss the extensive library of books on a multitude of topics. Designers must never miss the opportunity to be informed.

Figure 3-5.
Moreau title type treatment.

Cooper's fortitude and creative success with the R/Greenberg NYC studio put him in a position of power with the studio principals. The studio heads sent him to Los Angeles, California, the heart of the film industry. There he was instrumental in establishing the company's west coast studio operation. Eventually he would become one of the key players who would buy out R/Greenberg's west coast office entirely, shutting their doors and eliminating the Greenberg studio as a competitor in that part of the country. (The R/Greenberg studio has kept its base in NYC and continues to thrive as an innovative commercial studio.) In 1996 Cooper became a founding partner along with Peter Frankfurt and Chip Houghton in the top motion graphics and animation studio in the country by virtue of its pedigree and an A+ client list.

The company is called Imaginary Forces. The partners were the core creative group from R/Greenberg's west coast studio. Imaginary Forces became one of the premier motion graphics and digital animation studios in the country and continues to prosper—albeit in Cooper's absence. The success of Imaginary Forces is tied directly to the talent and vision of Cooper along with the founding partners mentioned previously. He left the studio on good terms shortly after taking a hiatus in 2000 to make his directorial debut on the film "Newport South." The film was a John Hughes production that Cooper is reticent about although he is clearly happy to have had the opportunity to move into directing.

Having sold his Imaginary Forces shares in 2003, Cooper set about the task of getting back to the work—listening to it. Realizing he had distanced himself from the creative process that he is passionate about in his role as principal—essentially spending much of his time engaging and managing people along with the day-to-day running of the

business—his decision to return to a hands-on approach to design is at the heart of his latest creative venture, Prologue. Prologue is Kyle Cooper. His creative journey has led him back to himself and what he believes he is happiest at—developing concepts, editing, and collaborating with an intimate circle of designers, animators, and editors. Building creative teams based on the complexity of each assignment allows him the freedom to keep his overhead to a manageable level while providing complete flexibility in choosing the talent that the project demands. He continues to work with long-time friend and associate Garson Yu, a connection from his days at Yale. However, the most important aspect of where he has positioned himself in relationship to his new studio is that he is not burdened with the need to engage a large staff that keeps him away from what he loves most—listening and responding to the work.

The definition that best describes his new studio is this one: An event or act that leads to something more important. This correctly characterizes Cooper's understanding of his place as a member of the film crew when he is working with a director in creating a film's opening credits. What is also evident in the name he chose for his studio is how it validates his approach and his respect for a film's director, typically the person who decides on the creative direction of the titles. Success breeds many things and in Cooper's case it has often put him in the untenable position of outshining the people he works for—the same people responsible for his opportunity as the film's title designer. He is uneasy with the kind of reviews that have at times placed his opening sequence design at odds with the movie itself. Critics have put the uniqueness of his title designs in a position that undercuts the films they introduce—placing Cooper between a rock and a hard place—praise at the expense of the hand that feeds him. *The New Yorker's* Anthony Lane proclaims on projects such as "Mission Impossible" that "Cooper's credits for the director Brian De Palma are so tense and sexy that you could leave the theatre immediately without suffering the letdown of the film itself." These kinds of reviews are hurtful and disturbing to Cooper. And in some cases have led directors to avoid working with him for fear that he may be out to undermine the film—thinking only of the impact of the credit design at the expense of the film. Collaboration is a critical part of the creative process in film title sequence design. This reality is cherished by Cooper and is evident in his seminal work, the title design for the movie "Seven."

The Handmade Title Sequence Design for the David Fincher Thriller "Seven"

The title design sequence of the 1996 film "Seven" was in fact the epitome of collaboration. The film's opening title sequence is a riveting visual and sound experience that moves viewers into the dark realm of the mind and soul of John Doe, the serial killer. I can think of few opening sequence designs that are as effective. Certainly the title treatment on the films "West Side Story" and "To Kill A Mocking Bird" were and still are standard bearers in this medium. Steven Spielberg's film "Catch Me If You Can" is a more recent success in this regard utilizing the hip stylized animation created by Dreamworks Studios. The film's title sequence pays homage to Bass by its use of graphic elements animated to underscore the coming narrative, also set to a great musical score. Bass also created the titles for "West Side Story." To be sure there were other standout designers in this often over-looked medium—Pablo Ferro, Robert Brownjohn and Maurice Binder are all standard bearers. Ferro's hand drawn titles for "Dr. Strangelove" inform the Fincher title sequence design. Brownjohn and Binder were responsible for the early James Bond titles

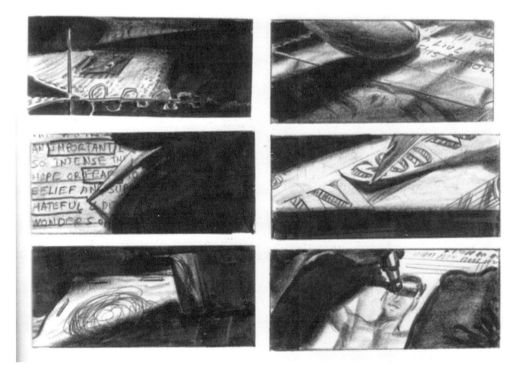

Figure 3-6.
Storyboard for "Seven" title sequence.

that effectively fused live action with the provocative use of typography in the opening title sequences, which set the stage for those films.

The ability to create imagery using type, form, and live action that are fused to the narrative in an arresting way is the essence of Cooper's title design. In the previously noted title designs, the soundtrack is a powerful ally to the visual experience taking place on the screen. The cathartic power of the musical score for the main title design can be a critical component of the effectiveness of the sequence. This is evident in the films mentioned above because each one successfully employs a powerful score, which reinforces the visual experience on the screen. However, it is Cooper's "Seven" that pushes that bar to greater heights with his application and layering of elements woven into the narrative of the film. It is clear why Fincher was successful as a music video director. The soundtrack's title score, performed by Nine Inch Nails, appears to have been created for this movie—and in a sense it was, given the remix of the song "Closer." The music is as sharp as the razor blade on the screen, which shows John Doe cutting the skin away from his fingertips. The beat and sparse lyrics help create the dark and foreboding world of the killer, who in effect has positioned himself as God. The visual jump-cuts and hand-drawn letterforms move the viewer through each layer of imagery by focusing on the obsessive nature of John Doe as he completes his manifesto, ever so precisely written and stored in the familiar notebooks we associate with innocence. As the pages turn on the screen, the music permeates the visuals like an open sore that festers to a crescendo within the tortured lyrics of the singer, "You get me closer to God." Cooper deftly uses still photos (shot by Melodie McDaniel with a Polaroid Land Camera and 665 film) that

have been purposely aged, distorted, and are sometimes out of focus, along with the live action shots of John Doe preparing his grim document (while having tea) with hand-drawn type, which was back lit and shot (Kodalithe). The original font was also distorted—Cooper's execution of the title design is evocative of his desire to bring to the film the human element—handmade design. He sold the idea to Fincher, a soft sell because the director and designer were on the same page from the outset. Cooper's creative process required that the title sequence be executed in such a way as to make what was a very precise and exacting design and execution methodology appear accidental. The concept of the manifesto notebooks used in the film came from the director David Fincher, who was influenced by Nine Inch Nails' music videos in the early 1990s. Cooper has the ability to create title sequence design for films that are visual fuses—lit by the designer himself, leading viewers to the narrative explosion that is to come. His work on "Seven" is a tribute to his diligence in creating the graphic realism of the optics used in developing his ideas. In David Fincher's thriller he achieved this by stepping inside the mind of the killer. At one point he retrieves human hairs from the shower stall (seen also in the opening sequence) to help enrich the visual madness and horror.

In many ways Cooper is a design aberration. He has one foot planted solidly in the modernist's approach to design, while simultaneously practicing on the fringes of postmodernism. He effectively uses elements of both in his approach to design—fragmentation, abstraction, and thoughtful juxtaposition. In viewing examples of his work, you can see his approach, particularly his clever and purposeful use of type as a moving image and the interplay of live action and narrative form.

Figure 3-7.
Spiderman title sequence.

Cooper's work on subsequent films, including both Spidermans show his diversity of thought and range. This suggests that he has not succumbed to a design formula, which often diminishes the work and relevance of successful designers. His work remains fresh. There are the designs that can be found in his work, particularly his animation, that are familiar, but how he integrates these forms with the overall optical on screen is what keeps them fresh.

The "Spiderman" titles created in collaboration with the director Sam Raimi are a testament to his diversity of style and creative approach. His use of type in the feature where letterforms are trapped like flies caught in Spiderman's web is used in the first feature. The strong use of Spiderman's web is carried over into the sequel. The screen is filled with a cascade of comic book images that build to reveal the all cap logo of Marvel Comics; these then move quickly to illustrated images from the first Spiderman film and are framed within the negative spaces of Spiderman's web. The black graphic web in effect becomes a frame much like that of the traditional comic book. This historic reference to the art of comics is

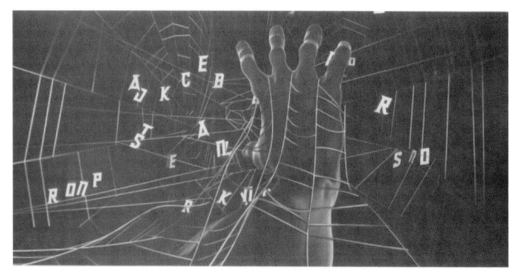

Figure 3-8.
Spiderman title sequence.

deliberate and very effective in setting the stage for the film sequel.

Cooper continues working in the commercial area of motion graphics for clients in film and television, where budgets can soar from $2000 to more than $50,000 for design development. However Cooper is in a position of some power as his reputation has placed him on the "A" list. This allows him the option to work on projects that interest him and ensures that he will be able to infuse the project with his design perspective while at the same time creating motion graphics that are designed in collaboration with the key principals on the project. When it comes to title design for feature films, that key figure is the director. Cooper is happiest participating in that collaboration as a member of the crew. It is in this setting that his best work is achieved. His creative process demands that he be true to his philosophy and approach to design. These requirements do not come at the expense of the project, rather, they ensure that Cooper will listen to the work, using a hands-on approach while working as hard and as thoughtfully as he can.

Chapter 3 Summary and Exercises: *Time Based Design Vernacular*

In 1995 Cooper single handedly rejuvenated the lost art of film sequence title design with a stunning visual manifestation on the big screen, creating a tabula rasa for motion graphics. This became the canvas for his visual prologue to feature films and motion graphics. In the design of the stills and live action title sequence for David Fincher's thriller "Seven" Cooper's creative process, while rooted in time-tested design principles, utilized a vernacular style to achieve the desired quality. It effectively created a dialogue between the brilliant killer and the theatre audience.

1. Designer's Keys to Success

Cooper's ability to stimulate audience perceptions is a result of his talent, education, and ability to move between a modernist and postmodernist approach to the manipulation of text, image, and sound. Juxtaposition, fragmentation, abstraction, and reality are used in provocative and thoughtful compositions on the screen. Collaboration between the designer and the film's director is the linchpin in his success as a feature film title sequence designer.

2. Open Thinking and Listening to the Work

Paul Rand's influence on Cooper's work is significant. His ascension as a designer working in the time-based medium of motion graphics is driven by the philosophy he adopted from Rand, which he described as listening to the work, and working as hard as you can. Invariably this is how the best solutions can be found.

3. Collaboration and the Informed Designer

The cornerstone of Cooper's creative process is staying open for ideas from all possible sources. In developing the look for the "Seven," Cooper focused on his collaboration with the director and crew in the use of jump-cuts and hand-drawn and back-lit letterforms. Polaroid shots from a Land camera using 665 film and notebook pages were used to create the contrast of obsession, innocence, and terror inherent to the film.

Design Exercise

Select one of the movies recently nominated for an Oscar for Best Picture of the Year. Research and begin developing an original title sequence using storyboards to illustrate the new creative direction for the feature film. (Initial ideas can be presented with black and grey markers on layout paper.)

- *Read the original script if available or watch the feature film from beginning to final closing credits.*
- *Review elements from the original title design that worked effectively; analyze weaknesses.*
- *Ensure that your design sets the table for the feature (the main course) using specific elements from the live action that enhance and dramatize your use of type, image, and sound in the visual prologue you create for the film.*

Professional Practice

Motion graphics requires expertise across several design and creative practices in the development of relevant, high-end, time-based concepts for a specialized and competitive market, which includes but is not limited to television broadcasting, filmmaking, trade/industrial videos, print, *and general consumer market advertising. Studios like Prologue, Imaginary Forces, Eyeball NYC, and R/Greenberg Associates excel as a result of their respective and unique creative approaches executed at the highest level and with the latest technology. Critically assess your studio practice and how it differs from the competition. Develop a promotional campaign for potential clients that dramatizes those unique attributes.*

- *Thoroughly research the competition; develop a creative approach that can be easily communicated to potential clients.*
- *Identify and highlight your approach to motion graphics, i.e., understanding and use of popular culture, music, and visual imagery, and explain why that approach makes your studio the best option for potential clients.*
- *Create a promotional CD and website that exemplifies your creative approach and clearly distinguishes your studio from the competitors.*

Graphic Archeology, DNA, and the Power of Design: Urban Poetry in the Built Environment

Deborah Sussman, Founding Partner of Sussman Prejza & Company, Inc.,

Culver City, California

Figure 4-1.
Deborah Sussman.

Introduction

Deborah Sussman's designs for the built environment have been called "urban poetry." From the award winning work on the 1984 Olympics to the Afrocentricity enhanced design of Newark, New Jersey's Performing Arts Center interiors, her eloquence of voice is the tapestry-laden backdrop of a design career that has propelled her to the front of the profession—across disciplines and to the height of excellence.

Deborah Sussman has created environmental graphics that gracefully and thoughtfully embrace the essence of the communities, cultures, and people who inhabit them. When she speaks of design and the Sussman/Prejza creative process she talks about "graphic archeology and discovering what is essentially the DNA of a project"—terminology not usually associated with the process of visual communications yet language that clearly validates the complexity of the Sussman/Prejza aesthetic and the seriousness of graphic application in the built environment. This process carefully engages all principals in what is often a complex working atmosphere. This requires the total "immersion of the creative team and following the threads of all the participants, including the people in the community, the client, the architects, the engineers, and the designers on the project." The beauty and wisdom of that statement is at the heart of Sussman's approach to graphic design in the built environment, where she continues to be recognized by her peers as perhaps the most relevant in this "layered and increasingly strategic business." It is that design sensibility that has infused the built environments and public spaces imbued with her distinct design aesthetic. Her creative leadership as the cofounder of Sussman/Prejza is the driving force behind the recognition of the studio and consistent success as a leader in the field.

The essence of the Sussman/Prejza design philosophy is the marriage of urban planning, architecture, landscape, and graphic design in the built environment. The wayfaring signage and overall graphic look of the 1984 Los Angeles Olympics was considered a seminal body of work for the firm, putting Sussman/Prejza on the design map. This opinion is also echoed by the national press in this quote from the January 1, 1990, Vol. 135, No. 1 issue of *Time* magazine: "Deborah Sussman's graphics and Jon Jerde's evanescent architecture for the games of the Los Angeles Olympics were homogenous, sunny, reassuring, *nice*. The color palette of the cardboard columns and fabric-covered fences was precisely of its time

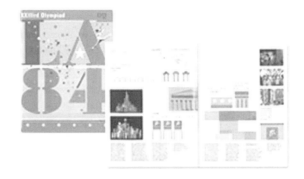

Figure 4-2.
1984 Olympic Games.

and place, beach-blanket postmodernism come to contemporary life."

Also mentioned in the article were other names of design significance—the Viet Nam Memorial, Apple Macintosh Computers, and the on-air graphics of MTV as well—very good company.

It is Sussman's ability to infuse the idiosyncratic details of the indigenous people and the areas in which they live, work, and play, that sets the design studio apart and places it ahead of the competition. It is a pacesetter in the extremely competitive field of environmental graphic design. The seamless application of type, form and space, and light and color, exemplifies the disciplined approach taken in solving distinctly different creative problems and is a classic example of the studio's relentless efforts in the creation of aesthetically

superior environments. The shimmering use of light created and applied to the interior space of the Seattle Opera house McCaw

Figure 4-3.
Olympic ceremonies.

Figure 4-4.
Olympic coliseum.

Other examples of the diversity of projects are the Euro Disney Theme Park located in France and Roppongi Hills located in Tokyo, Japan. These and other projects are the challenges that compel Sussman to keep moving forward and consequently maintain her position as a leader while she continues to lay creative claim to the landscapes of our urban centers. Her canvas remains the architectural structures and public spaces traversed daily by an unsuspecting public, representing environmental graphics at its best.

Overview

Sussman's career as a visual communicator spans three decades. Her approach and creative insight continues to be relevant amidst the superfluity of visual noise and look-at-me-designs that foster little in the minds of the viewer and often having little if any connection to the community it serves. Born with an intuitive and intellectual mind, the talent of Sussman has taken her far from her roots in Brooklyn, New York, and well beyond the hallowed halls of Bard College, Black Mountain, and the Chicago Art Institute where she was clearly a pacesetter among her peers. In fact she is more than that. Her story is that of the Renaissance man—except that this is the prolific career of a woman. Sussman's development and ascension to the top of her field was clearly ahead of its time—this would also manifest itself in the company she kept and in the educational and creative beginnings resonant in her early development as a visual thinker drawing on the ideals of mentors she had the privilege of engaging.

Hall, reinforced on each level and drawing on the mystical beauty of the northern lights, a phenomenon seen only in that part of the country—is a transcendent example of the Sussman/Prejza approach. It is a creatively strategic approach, which utilizes key elements of culture and environment in the creation of graphics and ambiance in the built environment.

Sussman/Prejza's client list stretches along the West Coast without ignoring the East Coast, and includes working internationally in urban settings in Europe and Japan. Often these diverse cultures are complex and not so easy to define as was the case in projects like the New Jersey Performing Arts Center located in Newark, New Jersey, a city with a scarred past. With this history always in the minds of the public, the project's potential impact and importance to the community was magnified. The center is the anchor of the "new" in Newark and has set the course for continued redevelopment—both structurally and in the minds of the community. The interiors created for that space are a celebration of the city's continued rise from the ashes of that history.

Multitalented and driven, Sussman was an aspiring young artist on a mission. She studied dance and performance with the innovative choreographer, Merce Cunningham, and musical composer, John Cage, at Black Mountain College. These two were innovators in their respective areas of artistic endeavor, and as such were also ahead of their time. Their approach to the development of young artists amidst the progressive environment of Black Mountain College had a positive impact on Sussman. The coeducational school was an experiment with its mission geared toward the cultivation of individuality and artistic development within a community setting where there was little separation between student, teacher, and family. One can easily reach the conclusion that these environments contributed greatly to Sussman's diversity of spirit and resolve to work as a designer where the complexity of each assignment reaches well beyond the creation of way-finding signage and provocative interior spaces. Success also rests on the ability of the designer to navigate the prevailing client culture and coordinate and cooperate with architects, engineers, and often donors with the collective goal of bringing the space to life.

Chosen from the relatively sheltered environment of the Art Institute of Chicago as a student by the famed first family of design, Charles and Ray Eames, Sussman left school early so she would make her mark under the tutelage of two of design's foremost practitioners. The importance of this experience on the young designer is evident and resonates in her tone as she speaks about the experience and opportunities it presented today, decades beyond the event. "Charles and Ray helped me to see the world more fully. I was sensitized to the importance of the environment, the people on the streets, not just from a distance—allowing yourself to experience the culture with an open curiosity."

During her tenure at the Eames' studio in California, Sussman was the essence of the yeoman assistant working on everything from catalogue covers to hand-drawn and inked artwork (using a ruling pen)—techniques and tools that predate the very concept of today's computer aided design. The Eames' experience left an indelible mark; the working relationship would last a decade, and prepare her for the opportunities that lay ahead. She was instilled with the confidence and determination to start her own design business. Charles and Ray Eames are considered two of the most important designers of the twentieth century. Creators of "The Chair of the Century," the couple worked in both architectural and industrial design environments. Their innovative use of molded plywood in various designs for furniture used in everyday life placed them among design's most revered thinkers. Eames furniture designs are still in production today—labeled as timeless beauties of practicality. With a background in architecture, Charles' and Ray's creation of the famous Case Study House in California, which utilizes the unique Eames approach of simple sophistication, remains a destination today for architects and designers from around the world for its innovative use of materials and

Figure 4-5.
Eames catalogue cover design.

its minimal approach to structural space. It is the tactile design legacy that embodies the essence of Charles and Ray Eames and their important place in design history and its future.

To refer to Charles and Ray Eames merely as mentors along the way is an understatement. Sussman adored them as sincere, caring people whose collective mission was to make the world a much more beautiful place to live, work, and play. Their advice to her when she reached the fork in the road that many progressive creative thinkers reach—whether to return to Chicago and continue her studies seems simplistic—but was actually very telling and insightful. "Why," they asked in essence, "would you take steps that move you in the opposite direction of your development?" She was clearly ahead of the curve and had eclipsed the program she was enrolled in at the time, intellectually and creatively. Her mentors were aware of this fact in selecting Sussman to work in their studio in the first place. These were individuals of vision and it was obvious to them that greater experiences were in store for Sussman—well beyond the city of Chicago. She never looked back.

Sussman would work in the Eames studio for more than ten years, with brief stops when she received the coveted Fulbright Scholarship to pursue the study of design abroad. This creative odyssey took her to Germany where she studied at the Hochschule fur Gestaltung. The design institution has a 200-year educational history and is one of Germany's most prestigious design universities with focus areas of media design, product design, and environmental design. Sussman remained in Europe for approximately three years. She would move on to France and Italy where she worked as a designer before returning to the United States and continuing her work at the Eames West Coast studio.

This period marked one of the most productive times for the Eames studio. Projects ran the gamut from creating exhibits for giants like IBM and the Ford Foundation along with government projects as far away as India. The range of specific projects and mediums was as diverse as the clients. The creation of showrooms for furniture designs as well as film and printed material were produced during this time. Sussman would work with Charles and Ray until deciding to go out on her own in the late 1960s. Deborah Sussman & Co. Inc. prospered for more than a decade before she created Sussman/Prejza with Paul Prejza, whom she married in 1980. Paul Prejza's background in urban planning and architecture made for the ideal partnership. A graduate of Penn State's College of Art and Architecture his shared goal of reaching beyond brand identity for his clients, specifically municipalities, was to effectively link these urban enclaves with imagery that created the specific "feel" of its location and culture. Working as an engineering planner on the Apollo moon-shot program at North American Aviation, Prejza brought a technical focus to the design team. This approach provided the leadership that encompasses analysis of complex creative assignments in the built environment where the highest level of execution and performance is demanded. While the focus of this chapter is specifically on the outstanding career of his wife and business partner Sussman, the growth and prosperity of Sussman/Prejza benefits immensely from the working relationship of both its principal founders, validating Prejza's contribution to the success of the firm from its inception.

Northern Lights: The Distinctive DNA of Seattle's McCaw Opera House

When Sussman discussed the work Sussman/Prejza completed on the newly renovated McCaw Opera House, her eyes

Figure 4-6.
McCaw Opera House lobby.

sparkle with a glint of the famous northern lights and shimmering reflections associated with that part of the country. Her anthropological approach to defining the look of the interior spaces was driven by the details of the environment and the public that would utilize the space.

The visual elements created on each level of the interior space underscore the architectural design of the exterior as well. This quote by Sussman in the June 28, 2003 *Seattle Times* article written by Melinda Bargreen speaks to her implicit attention and appreciation of detail: "I responded to the quality of light in Seattle—the lobby faces the sunset. I thought about those colors, and also about the time I flew through the Aurora Borealis. Northern colors are not all grey and brown. The vertical walls of the lobbies have striations or bands of colors that you get in the sunset—and the shimmer of rain and wet surfaces are reflected in the design of the carpet and the terrazzo tile. The carpet, which is a dark wine, carries out in its design the idea of water droplets and rivulets after the rain. The upper lobby walls are also reflective, changing colors as you walk by."

The previous statement speaks clearly to the importance of "seeing" in Sussman's creative process—paying attention to the environment, its beauty, and its visual intrigue—and the infusion of these nuances into the design. This is one of the great strengths of the firm and its approach in creating designs for the built environment—celebrating these contiguous elements—as one great space, or better stated—canvas.

Sussman created and worked from a palate that included more 100 wall finishes, more than 90 colors of paint, and various other fabrics, plaster, and metal leaf. Part of the challenge of a firm like Sussman/Prejza is working with a complex client and project structure, which in this case included the architects, the artistic director of the opera

Figure 4-7.
McCaw Opera House hall.

Figure 4-9.
McCaw Opera House lobby. (See also Figure (xxi) in the color insert.)

Figure 4-8.
McCaw Opera House trustees room.

and Pacific Northwest Ballet, and the Seattle Center administrators as well as key donor Susan Brotman, whose name is on the auditorium. The review and approval process for this project was extremely extensive. This is not out of step with most creative reviews; however, given the number of critical people brought into the process, the road to final approval was not without the bumps in the road typically associated with a complex project. The underlying and most important aspect of the project was that Sussman and the client "remained for the most part on the same page." It appears that the review process allowed for openness and thinking the ideas through based on comments from Sussman. The willingness of such a complex group of clients to trust her judgment is a testament to the firm's successful and international reputation. I also suspect that Sussman's eclectic background and vigor made her a formidable motivator and presenter of the ideas that may have been out of the norm—as were many of her choices for the McCaw. She herself states that her selections for the space and environment were "definitely a stretch."

The overall renovation and expansion of the Opera House by the architectural firm LMN cost $127 million. Another key to the look of the renovated McCaw are the dramatic scrims at the entry of the Hall's Kreielsheimer Promenade created by New York-based artist Leni Schwendinger. There are nine, 30-foot-tall mesh scrims in all, suspended between McCaw Hall and the old Exhibition Hall. The phenomenon of the visual experience for guests entering the hall and passersby on the street is again that of the lights in the Aurora Borealis, hovering between the glass walls of the opera house and the Phelps Center. This design underscores and coordinates with the color selection that Sussman applied to the interior spaces.

Many of the nuanced visual goals articulated by Sussman find themselves dramatically mirrored in the illuminated scrims designed by Schwendinger, from the rain soaked surface of the city to the multitude of colors that are associated with the northern lights. Created to be an additional theatrical experience, the scrims are programmed with an astrological clock

which changes during the evening hours—setting the stage for a visual experience for Seattle's citizens.

NJPAC | New Jersey's Performing Arts Centre, Newark, New Jersey

Several years ago Mayor Sharpe James renamed the maligned municipality that lives in the shadow of Manhattan's sophisticated vivacity—New Jersey's Renaissance City. He was expressing the hope that remained in his heart during his tenure as mayor that soon Newark's resurgence as a safe and prosperous city in which to live, work, and play would be complete. This hope is shared by many familiar with the legacy of a metropolis still healing from the political and social turmoil that fueled the civil disturbances of the 1967 riots and the corruption that preceded it and has been embraced by newly appointed Mayor Cory Booker.

The citizenry of this increasingly diverse population is spread across its north, south, central, east, and west wards. Today it is populated by 4.5 million people, more than half the population of the entire state. Settled originally in 1666, its skyline borders the west bank of the once fertile Passaic River, which roped itself through the industries that had helped the city keep pace with other major centers of commerce and productivity. Breweries and factories covered the areas from McCarter Highway to Frelinghuysen Avenue. Although those industries are long gone, this distinct area of the city is home to one of the most important performing and cultural arts structure built in an urban center. The goal is to anchor this former city of blight to the hope and prosperity that has eluded it for decades. The road to fruition was long and began in the mid-1980s, when the foresight of Governor Thomas Kean and Mayor Sharpe James was essential to the planning and creation of this new performing arts center. Key to the creation of NJPAC was the remarkable public-private partnership, which included donors, business from the private sector as well as state and local government.

Moving north from the heart of downtown or "down neck" and approaching the complex from either McCarter highway or Broad Street via Military Park, the structure commands your attention and the dramatic façade of the edifice embraces you albeit with the suffocating arms of a love-starved great aunt. In this, the overall architectural design of NJPAC is arguably too visually aggressive; however, the beauty of the logotype design, created by Sussman/Prejza quickly emerges and moves you beyond the labyrinth of steel piping that is the outside layer of the structure.

The logo contrasts effectively with the marvelous federal style architecture that is a historic reflection of buildings located throughout the city of Newark. The font design is a balanced application of typographic form, space, and image. Modified from the classic typeface Centaur, designed by Master Typographer Bruce Rogers, the logotype is set low and wide in cap height applications, the letterforms appear as if standing on stage with a solitary spotlight in the form of an oval shown directly on the letter "A." The letterform breaks into the negative space of the solid oval shape. The top edge of the oval spotlight serves as the crossbar of the letter "A" as well—the obvious emphasis here being on the word "Arts" at New Jersey's newest and most important performance venue.

Figure 4-10.
NJPAC logo. (See also Figure (xxii) in the color insert.)

Figure 4-11.
NJPAC carpet pattern design.

This clever usage of font and form gives the logo a memorable graphic gestalt. The open-spaced kerning of the logotype design gives the structure a sense of dimension as well as visual presence and sophistication. The subtle visual grace to the unit structure is appropriate and deserved. The oval spotlight, which appears as a burnt orange, contrasts with the green of the letterforms and is repeated throughout the interior space and onto way-finding systems and patterns guiding and welcoming visitors to the venue. The huge signage atop the main entrance is executed in metal—again the power of visual contrast is at work and the letterforms contrast with the federal red brick of the exterior of the structure.

As with the McCaw, Sussman/Prejza is at the top of their game when it comes to the nuance and power of interior design that is rooted in the diversity and culture of the community it will serve. The interior space of NJPAC pays visual homage to the people that make up the city, with ethnic nods to Africa and Latin America. The colors and patterns reach out to those collective cultures mentioned and embrace them visually, from carpeting to signage and back outside to the banners that surround the complex. Sussman/Prejza's prowess with the application of the "right" color pallet is once again a visual realization of Sussman's anthropological approach to designing an environment that is reflective of the people of the city of Newark.

Figure 4-12.
NJPAC ID sign.

Figure 4-13.
NJPAC directional.

The City of Santa Monica

One consistent element streaming through the award winning work created for the 1984 Olympics was Sussman/Prejza's bold application of color—from the way-finding systems to the ornamentation-adorned stadium and event graphics throughout the site. When the Sussman/Prejza design team took on the challenge of developing the identity for the City of Santa Monica, austere graphic shapes and the rich use of color set the tone for visuals that rejoice in the stylish and upscale atmosphere.

Once known primarily as a haven for the Angelinos of privilege the city's historic pier, built in 1908, boasts breathtaking panoramic views of the great Pacific and is anchored by a rustic small-town amusement center, replete with ferris wheel, roller coaster, and a carousel with hand-carved ponies. It is this small slice of Americana that is the spirit of the graphic system developed by Sussman/Prejza for the thriving municipality. The goal of the revitalization of the sleeping coastal city of Santa Monica was met head-on with the power of a bus—actually a big blue bus

and the aforementioned visual approach of Sussman/Prejza. The simplified sun-drenched winding road and hill top shapes are the main visual continuum on most of the applications for the town. It appears as a

Figure 4-14.
NJPAC banner 2. (See also Figure (xxiii) in the color insert.)

Figure 4-15.
Santa Monica logo.

and a cerulean blue indicative of the sweeping and often cloudless skies. These colors were refined from the original seal of the city, which remained in the mindset of the client and is a cherished relic of the municipality's past.

The Sussman/Prejza mark successfully rejuvenates the seal while holding on to the past. The new identity developed by the firm included guidelines for all applications. The spirit of the design approach is a calculated, light-hearted direction bordering on whimsy. The clever use of the throw-back Bodoni-inspired serif font for the city nomenclature set against the clean contrast of the classic sans serif font precludes the overspill into the juvenile realm. This visual contrast of design elements, fonts, form, and space, is tried and true—calling the viewer's attention to it in a friendly and sophisticated way.

Sussman/Prejza's consistent use of color in the creation of environmental design elements for the city of Santa Monica keeps pace with the high marks that were set in the development of the 1984 Olympic Village project, which put the firm on the world

roundel on each street name, just above the rectangle and to the right, as well as on municipal vehicles for service and commuters. Collateral materials for the town, such as maps and tourism pamphlets utilize these graphic elements as well.

One of the dominant colors used is a golden yellow, a color reflective of the sun as it bathes light and warmth on the beautiful beaches and the coastal highway nearby,

Figure 4-16.
Welcome sign. (See also Figure (xxiv) in the color insert.)

Figure 4-17.
Street sign 2.

stage three decades ago. Sussman developed the identity of the city with her anthropological approach to problem solving—searching out the essence of the community and its indigenous inhabitants by using the "cultural DNA" of the built environment. To this end, Sussman effectively celebrates the "flavor" she creates through design, which for Santa Monica is an interpretive mix of shorelines, cotton candy, peanuts, and strawberries—dipped in melted chocolate.

Hancock Park, Los Angeles, California: A Natural Oasis in the Heart of the City

The development of the Hancock Park project comprised landscape, environmental, lighting, and architectural design. The award-winning project was implemented in the setting of a natural historic site located in the heart of Los Angeles. As the location of the LA County Museum of Art and the Page Museum, the park combines science and art. The way-finding systems developed by the

studio enhance the natural setting with a color pallet utilizing earth tones of golden yellow and burnt orange.

The thoughtful use of graphic elements for directional signage and site maps are engaging and easily followed by the public. The viewing stations are more than just practical. They are constructed at a level accessible to small children and to adults. The signage in and around the site is visually reflective of the fossils and other natural shapes

Figure 4-18.
Big blue bus.

Figure 4-19.
Parking sign.

Figure 4-21.
Street sign 4.

and elements found in the park. The fact that this natural oasis location is just minutes from downtown LA, is an accomplishment in and of itself. Classic and legible typography

is used, contrasting the graceful serif font of the Hancock Park logo with a utilitarian application of a sans serif face.

The subdued and landscaped park is a respite from the busy and noisy city that is its home. The visuals developed by Sussman/Prejza for this project blend seamlessly with the environment and complement the natural and artistic experiences of each visitor.

Euro Disney

Sussman/Prejza's environmental graphic solutions for Walt Disney's theme park in Paris, France utilizes a bright color pallet created specifically for Disney and consistent with the Sussman/Prejza approach to the development of clean and elegant graphic signage systems. The creation of the theme

Figure 4-20.
Directory.

Figure 4-22.
Hancock Park viewing station A. (See also Figure (xxv) in the color insert.)

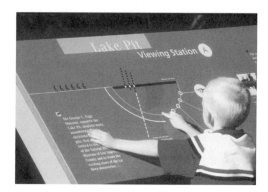

Figure 4-23.
Hancock Park fence mounted lake signage.

park's logotype and the multiplicity of signage way-finding systems and graphic applications follow visual notes reflective of the original Olympic Village project. It is impossible for one to disassociate the lively and familiar iconography associated with the legendary

Figure 4-24.
Hancock major ID sign.

Figure 4-25.
Hancock directional sign.

Disney franchise nor should we, as these images have become a part of the American experience, from the environmental design elements created for the project. Transplanting this theme park experience to Paris, France, would be quite a challenge for the developers and initially opened to lukewarm reviews by the press. This was

Figure 4-26.
Hancock Park "do not pick flowers" sign.

Figure 4-27.
Euro Disney logo.

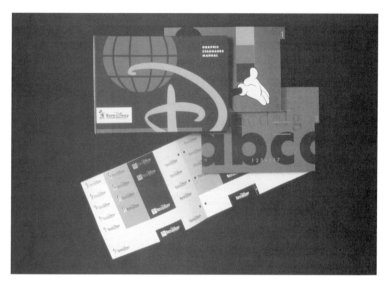

Figure 4-28.
Euro Disney graphic
guidelines.

systems created for the park do not break new ground, they are consistently implemented, reinforcing the bright and enchanting atmosphere of the park. Again the sophisticated use of type and graphic elements are effectively applied to street signs, buses, and site maps in addition to developing the graphic standards manual for the usage and application of the official logotype.

Sussman has been developing visual communications systems for the built environment for more than three decades—her rise to the top of the profession is due to talent, and hard and smart work. In the creation of award-winning designs for a diverse range of international clients, Sussman/Prejza has established itself as one of the foremost practitioners of this highly specialized area of design practice. When speaking about a lifetime of work and accomplishment in designing for the built environment, Sussman exhibits her passion for design and a profound love of its practice. She continues to apply the lessons learned as a young designer

not the fate of the environmental graphics developed by Sussman/Prejza. While the applications of signage and way-finding

Figure 4-29.
Euro Disney shuttle bus.

working with Charles and Ray Eames: that good design is about seeing and experiencing the world around you in an informative way.

Figure 4-30.
Euro Disney bus stop.

Chapter 4 Summary and Exercises: *The DNA of Environmental Graphics*

In 1984 Sussman/Prejza established itself as a leader in the practice of graphic design for the built environment with its seminal design of architectural structures and way-finding systems for the Olympics as well as servicing a long list of international clients. With the leadership of Sussman, whose designs have been referred to as "urban poetry," the firm established itself through its distinctive creative process, which embraces communities and the people and cultures who shape them—from Newark's Performing Arts Center to the thriving shopping and business district of Tokyo's Roppongi Hills.

1. Designer's Keys to Success

Sussman's ability to define and then implement what she refers to as the "DNA" of a project into design solutions for Sussman/Prejza's clients is the definitive creative approach of the studio she heads.

Figure 4-31.
Euro Disney directional signage.

Her creative process requires the total immersion of the creative team, including the client, the community, the architects, the engineers, and the designers. The philosophy of the firm is the cohesive marriage of urban planning, architecture, landscape, and graphic design for the built environment.

2. Urban Poetry and Visual Harmony
Named in 1990 by Time magazine as one of the best designs of the decade (1984 Olympics), Sussman's environmental graphics were heralded as "harmonious and reassuring. The color palette was specific to its time and place as postmodernism met the contemporary lifestyle of LA." Sussman/Prejza's use of color continues to be a strength resonating in the structures and way-finding systems designed by the studio today.

3. The Transcendent use of Form and Space, and Light and Color
Creating environments that are aesthetically superior such as the shimmering use of light and color applied to the Seattle Opera House (McCaw Hall), Sussman replicated the beauty of Seattle's northern lights, illustrating the depth of the Sussman/Prejza creative process. This process embraces the existing environment, the community, and its people.

Design Exercise
Select a municipality, which historically has image problems in the public mindset. Redesign its identity, street signage, and one area of public transportation for a specific group of users, for example, senior citizens, the physically challenged, etc.

- Research the history of the city, as well as present perceptions of it as a viable community in which to live, work, and play; analyze all relevant issues.
- Beginning with the graphic identity of the city, utilize specific elements reflective of the history, diversity, and culture in the

design of the new image—essentially defining the DNA of the municipality and infusing that essence into the visual communications and way-finding systems of the city.

- Apply elements of the newly designed identity to all applicable items creating a visual link between the culture and the people throughout the municipality.

Professional Practice
One of the key elements of environmental graphics is effective collaboration and coordination between the client, the community, the architects, the engineers, and the designers in developing effective solutions. Many studios are multidisciplinary and serve clients in the area of environmental graphics, corporate identity, way-finding and signage systems, interior design, and architectural color and finishes.

- Thoroughly research current business opportunities in local municipalities; urban centers are actively seeking partnerships in the rejuvenation and restoration of public and residential spaces. Develop a creative approach that emphasizes your commitment to their local efforts for improvement and visibility through your knowledge of the community and its current culture and population.
- Highlight your creative process and understanding of the history and popular culture specific to that urban setting; present the focus of your professional practice in a context with clear linkages to the community.
- Pitch a community-based project designed to enhance the city's image through the strategic use of environmental graphics, i.e., marking of historical and relevant sites throughout the community. Develop a campaign targeting the clean-up of empty lots for use as gardens and recreational spaces, etc.

Urban, Environmental, and Architectural Design: Historical Contrast in a Cultural Context

Mi Qiu's Modern Art Workshop
Shanghai, People's Republic of China

Briefs Highlighted: Divine Existence
The Beijing Project Nan Xingcang

Clients: China Life Insurance Co., Ltd
H. K. I. Development

Figure 5-1.
Mi Qiu.

Introduction

Historically the nature of cities, particularly since the invention of the automobile, have been that of a living organism in divergence with its own inevitable demise and regeneration. This struggle is archetypical of technology's effect on society, particularly in urban settings. At the turn of the nineteenth century the horseless carriage gave the citizenry the ability to move beyond the confines of the metropolis and set the tone for today's sprawl, which is endemic to societies in developmental overdrive. This situation is not unique to western culture. In China, the growth rate in major cities continues to increase at an astonishing pace. In Shanghai, often referred to as "The Paris of the Far East," new building and construction projects continue to move the city forward, or more aptly put, upward. In both business and residential settings, the buildings move closer and closer to the clouds. The overall design of these structures and public spaces reveals a hodgepodge of architectural styles, except The Bund where the imprint of nineteenth century British architecture remains. In China the migratory trend is reversed—people from the countryside are moving in waves to the cities. This

phenomenon is much like America's black migration of the mid-twentieth century, when multitudes of African Americans moved from the south to the northeast for better job opportunities in factories and businesses.

In Shenzhen alone, the population has grown to more than seven million over the past several years. That city is a new creation and does not have the same dilemma regarding the protection of ancient structures as does China's flagship cities such as Shanghai and the country's capital city, Beijing. The government has made a concerted effort to retain and restore where possible some of country's original urban structures that date to the Yuan, Ming, and Qing Dynasties. These buildings vary from royal granaries and warehouses to simple residential dwellings. With half of all the concrete poured in the world being poured in mainland China and as much as seventy-five percent of the world's steel being consumed, it is not surprising that China's ancient and culturally significant buildings are in peril. They could disappear at an alarming rate without creative ways of incorporating new designs with the relevant ancient structures that should not be destroyed. This trend in new construction speaks to the great economic prosperity that China is beginning to experience as a major player in the global economy.

The economic pace in China is such that the government is considering slowing growth for a more stable market as China continues to emerge as a leader on the global financial markets. These are all compelling issues that Mi Qui has successfully navigated—making it his personal mission to create new structures and public spaces that include the cultural pearls of China's magnificent history while providing the community with urban, environmental, and architectural designs that celebrate the past, present, and future. He has managed to do this at a high aesthetic level—remaining true to his artistic sensibilities and goals in designing spaces for living, working, and enjoying life. He has accomplished this while embarking on a creative crusade encompassing architecture, environmental, and landscape design as well as fine art, which includes a diversity of mediums such as installation, performance, sculpture, and painting. More compelling than the mix itself is the great success of these mediums as a means to a creative end that uplifts the spirit and lifestyle of the public privileged to live, work, and play in these spaces.

Overview

Mi Qiu was born in Shanghai, China, in 1960 during the height of Chairman Mao Zedong's leadership. He along with his mother, father, and four siblings all endured the upheaval of China's notorious Cultural Revolution. He explained, "we all lived and slept in one room during that period—all nine of us. My grandparents were also with us." China's Cultural Revolution lasted ten years, from 1966 to 1976 representing one of the most tragic and difficult periods in the history of the country. Facilitated by Chairman Mao Zedong in a political attempted to maintain a firm communist structure of the country while ensuring the desired transition of power— effectively set the progress of this great society backwards. These difficult circumstances did little to discourage Mi Qiu's educational and artistic ambitions, partly due to the relationships that he had and continues to have with his family. Mi Qiu's grandfather was a Taoist priest at Xuanmiao Taoist Temple in Suzhou. His relationship with his grandparents was particularly close. The memory of his grandmother's stories told to him as a child and the spirituality and wisdom of his grandfather would resonate throughout his life and become a beacon of light central to his creative process.

Advancing successfully through school and completing his college studies, Mi Qiu graduated from Tongji University, in Shanghai, with a degree in architecture. He immediately went to work at China's City Design and Research Center from 1982 through 1987, where he participated in reconstruction projects in twenty-four historical cities as well as forty-four government-approved conserved landscape locations throughout China. Most of the residential buildings in China were of the same style and appearance, and although Mi Qiu had majored and excelled in architecture and was successful in his position at the research center, it was difficult for him to put forth his ideas during this period. His was a restless and hungry creative soul. The crucial phase of true growth and experimentation was yet to come. In 1987 he would embark upon what was to become his artistic pilgrimage, and while this journey would take him beyond the great walls of the People's Republic of China for the first time in his life, experiencing western modernism and culture first-hand, his quest for artistic exploration, realization and enlightenment would propel him forward as a visual artist.

Emigrating to Europe in 1987, his evolution as a contemporary visual artist found fertile soil for his distinctive approaches to creative expression. Since that time he has participated in more than thirty exhibitions around the world. Painting, sculpture, interior design, landscape design, urban planning, and performance art are all areas that Mi Qiu has realized successfully as an artist. His experiences as an exhibiting fine artist set him apart from many in his field, and his desire to uplift the human condition through his work is genuine and reflected in designs that are now scattered throughout China's major cities.

Mi Qiu is one of China's first exhibiting contemporary artists to break through that historically restrictive country. It is easy to conclude in reviewing his work and growth over the last seventeen years that Mi Qiu's work ethic is as prolific as his artistic diversity of mediums. In 1993, he established the Europe-China Culture Exchange Project (ECCEP). In 1995, he served as a researcher at Henie-Onstad Art Center. In 1996, he established the Mi Qiu Modern Art Workshop in Shanghai, where more than thirty art exchange projects were conceived, created, and exhibited. In 1998, he received the prestigious "Movado Outstanding Contribution to Art Award." That same year, he created the "ArTech" project, which successfully combines the pursuit of artistic creativity and imagination as part of the conceptual process in urban planning and design. During this time period Mi Qiu focused on the conception of the visual synchronization between environmental design and sculpture.

Among the projects he created that are a testament to his visual approach to environmental design is the eighty-meter glass bridge, which spans the river in the Shenzhen Sculpture Park. Mi Qiu carved text directly onto the glass panels that run the length of the bridge that speak to the future and survival of the human race. He takes the opportunity for self-expression as it relates to the benefit of his fellow man. As people traverse the glass bridge, they are enraptured with a sense of peacefulness while experiencing the graceful memory of history, which flows in the river below it.

As the decade ended and moved into the new millennium, Mi Qiu continued to conceive of structures in public spaces that heralded the mind-set of China's new generation—a generation that many in the media have dubbed him the leader of. Mi Qiu's popular status in Shanghai keeps him in public view, as he is a celebrity in a country that embraces its own—particularly the ones who return home. This public appreciation is validated by his genuine love for his country and home city. He belongs to the first generation of Chinese artists widely exposed to western modernism. Beyond exposure to this different point of view is the understanding of it and its subsequent infusion and reinterpretation through the creative minds of China's contemporary visual artists of which Mi Qiu is one of its innovators and leaders.

Mi Qiu's work continues to embrace the human spirit, individuality, and culture of China—the country that celebrates and nurtures it; he defies any fixed label. He has successfully and purposefully blurred the lines of abstract painting, sculpture,

Figure 5-2.
Fax action detail.

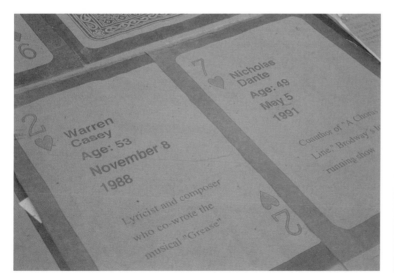

Figure 5-3.
Fax action AIDS Awareness Project 2000.

performance art, architecture, and environmental design as elements of each find their way into the collective manifestations of his artistic expression. The projects that follow represent the complexities of a creative process that embraces China's ancient history, its present and future, and the soul of its people.

Ancient Passages in Contemporary Spaces | Divine Existence

As you walk through Shanghai's Ningshou Plaza located at the corner of Huaihai and Renmin roads, the smell of change lingers in the air. Shanghai moves forward like a force of nature—occasionally without regard for what may be in its path. The rubble and debris from demolished residences and businesses keep that smell fresh. This area is the focal point of the reconstruction taking place in the old neighborhoods of Shanghai—and in this chapter the focus and fruition of a creative, cultural, and historic design opportunity for Mi Qiu's artistic vision. Entire communities are being reborn as the rapid development of Shanghai continues.

The residential community directly across the street from the China Life Insurance Co., Ltd. (CLIC) was being demolished and replaced with new apartments, condominiums, and shops that were more reflective of the new wave of young, sophisticated, and upwardly mobile Shanghai residents who would patronize and occupy them. Standing at the entryway of this area was a regal and neglected archway, which was at the time set to be demolished with the other dwellings behind it. The two-hundred-year-old archway dated back to the Qing Dynasty and was called Ningbo Assembly Hall.

The archway is a formidable red brick structure with Chinese characters in relief directly above its cement-bordered pathway.

It is crowned with distinct concrete detailing at four points, and a higher point, the fifth, sitting approximately twenty-five meters from the ground. Mi Qiu was hired to design architectural structures and environmental sculptures that would enhance and unite the buildings that comprised the CLIC corporate headquarters located directly across the street from the archway and the old neighborhood. From the moment Mi Qiu realized the neglected archway was available, and understanding the historical nature of the structure and the conceptual possibilities inherent in its juxtaposition, he set out to save the structure and to make it the critical design element of the project. He wanted to incorporate China's past, present, and future. To do this, the archway would be moved.

Mi Qiu's influences in his approach to and realization of design are as eclectic as his visual arts background suggests—from his experiences collaborating with the artists Cristo and Jeanne Claude during his nine-year exploration in Europe to his reverence for the two Franks—Wright and Gehry. He revered Wright for the sublime majesty of his approach to architecture and his influence on modern architecture in the twentieth century. And he appreciated Gehry for his ability and courage to boldly reinvent himself by utilizing materials previously unseen in the manifestations of modern architectural structures, essentially turning the structures inside-out.

The post-modernism of Phillip Johnson also resonates in the creative approach of Mi Qiu. In much the same way that Phillip Johnson spoke of structures having the possibility to be conduits or clearing houses for the development of ideas (speaking specifically of the Glass House in New Canaan, Connecticut), Mi Qiu, through the clever juxtaposition of elements and the contrasting visual effects of materials within the structures he designs, gives the public the

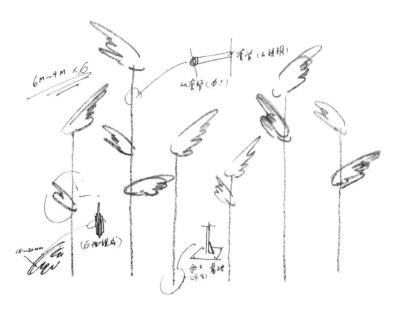

Figure 5-5.
Way of Existence wing detail. (See also Figure (xxvi) in the color insert.)

Figure 5-4.
Thumbnail sketch for "Way of Existence" wings.

opportunity to experience his ideals. This he does with the unabashed joy of a child and the sophistication of the talented and progressive artist that he has become. He has spent significant time reviewing and studying

the work ethic and results of others as he has honed his approach to design. He has room for the visual arts and also the written word, as it has found its way into his environmental designs—etched into a glass bridge in a sculpture garden in Shenzen called Way of Existence. Here he uses the actual written

Figure 5-6.
Way of Existence etched glass walkway detail. (See also Figure (xxvii) in the color insert.)

Figure 5-8.
Way of Existence etched
glass walkway and wings.

Figure 5-7.
Way of Existence etched glass walkway and wings.
(See also Figure (xxviii) in the color insert.)

words on one panel moving to their digital equivalent on the others—skillfully and playfully acknowledging technology in the process of the experience—telling the story of man's existence as experienced and seen through his eyes. This would seem to be at odds with a design background in architecture where "form follows function" is the prevailing rule.

In design, where the intentional ordering of the natural or immediate environment is expected (be it three-dimensional structures or images on the printed page), these experiences can be nondescript and taken for granted. The CLIC structures created by Mi Qiu engage the public in a way that while unobtrusive cannot be ignored, or remain un-experienced.

The CLIC commissioned the project to Mi Qiu and his firm exclusively. Often at the inception of a job, particularly in China, the

selection process for a project of this size involves several architectural and environmental design firms submitting proposals for review and approval by the client and the local government. Mi Qiu's success and reputation gave him a distinct advantage, and his modesty belies a creative focus and competitive nature that has put him in an enviable position. He is often

Figure 5-9.
Way of Existence aerial
view of sculpture garden.

Figure 5-10.
Thumbnail sketch for
Divine Existence.

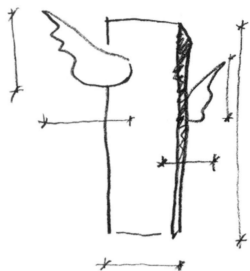

Figure 5-12.
Thumbnail sketch for Divine Existence.

called upon to take on projects after the client has reviewed and rejected the submissions of other design firms.

The CLIC plaza is in an area situated between two buildings. The total space is approximately 200 square meters with an additional 100 meters in height. The new environmental structures contrast historical and architectural elements within contemporary design structures in urban environments and public spaces. The archway was moved twenty-six meters from its original site—a perilous task orchestrated by engineers and construction crews under the guidance and supervision of Mi Qiu. It is

possibly the country's second ancient historical structure to be relocated rather than destroyed (the first being the astronomy tower, which is also located in Shanghai). Mi Qiu also incorporated a glass wall (eighteen meters long by two and four-tenths meters tall), which streams between the archway and the new plaza to visually demonstrate the passage of information and time from one structure or period to the next. Mi Qiu created abstract drawings in ink-washed lines, which were then carved using optical fibers directly onto the ten-cm thick glass wall. These remind us of the traditional ink-washed paintings that inform his approach to the drawings on the wall and, as a medium, are embedded in China's visual arts culture and Mi Qiu's own artistic exploration during the 1980s. When the structures are lit, the glass wall contrasts with the warm colored archway and forms a memorial venue representing Shanghai's past, present, and future. The line drawings flow playfully and dramatically on the glass wall through the archway, suggesting an infinite journey from the past and into tomorrow. Mi Qiu's coupling of contrasting forms through a

Figure 5-11.
Thumbnail sketch for
Divine Existence.

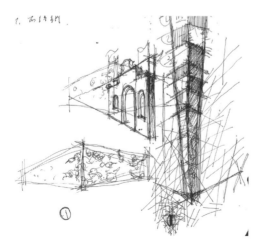

Figure (i)

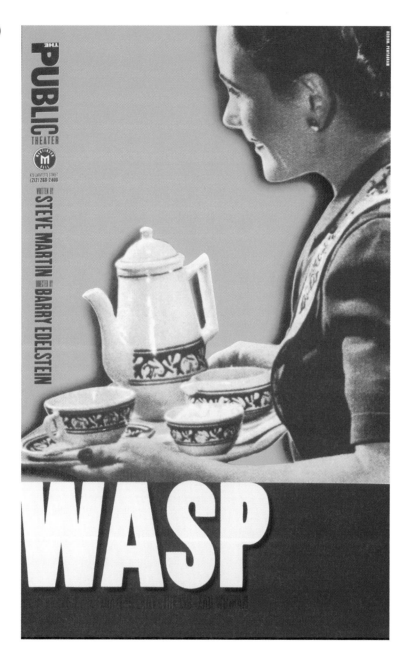

Figure (ii)

Figure (iii)

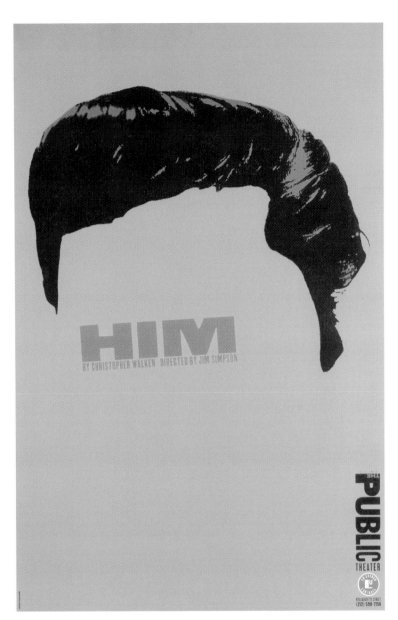

Figure (iv)

Figure (v)

Figure (vi)

30 PERFORMANCES ONLY

MARCH 4 - APRIL 13

BALLET·TECH

A NEW COMPANY FROM ELIOT FELD

CALL JOYCECHARGE 212-242-0800

The Joyce Theater / Chelsea 175 Eighth Ave. at 19th. St.

Figure (vii)

Figure (viii)

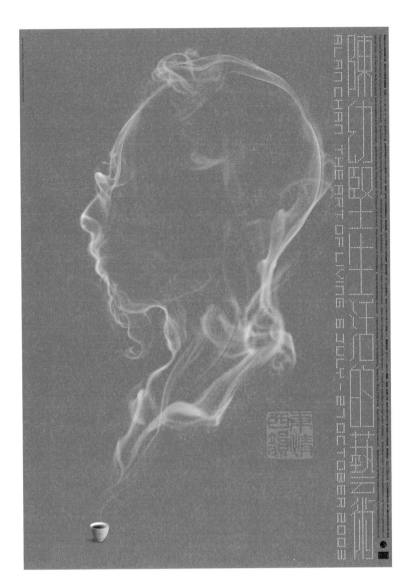

Figure (ix)

Figure (x)

Figure (xi)

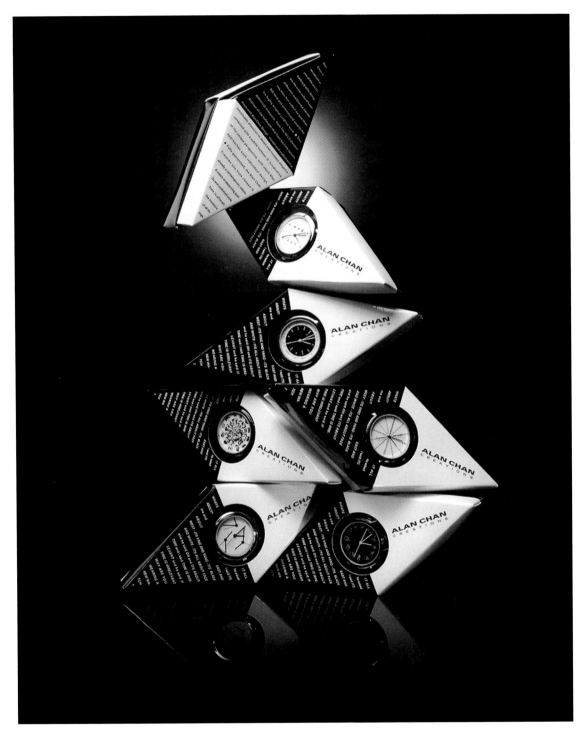

Figure (xii)

Figure (xiii)

Figure (xiv)

Figure (xv)

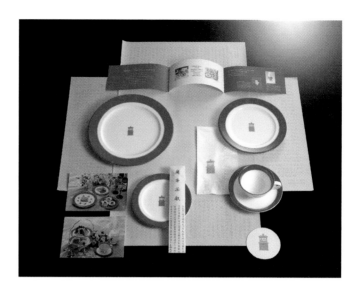

Figure (xvi)

Figure (xvii)

Figure (xviii)

Figure (xix)

Figure (xx)

Figure (xxi)

Figure (xxii)

Figure (xxiii)

Figure (xxiv)

Figure (xxv)

Figure (xxvi)

Figure (xxvii)

Figure (xxviii)

Figure (xxix)

Figure (xxx)

Figure (xxxi)

Figure (xxxii)

Figure (xxxiii)

Figure (xxxiv)

Figure (xxxv)

Figure (xxxvi)

Figure (xxxviii)

Figure (xxxvii)

Figure (xxxix)

Figure (xl)

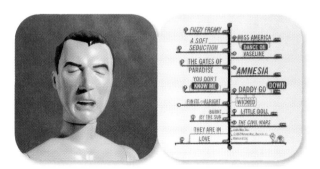

Figure (xli)

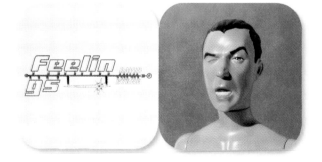

Figure (xlii)

C-30

Figure (xliii)

Figure (xliv)

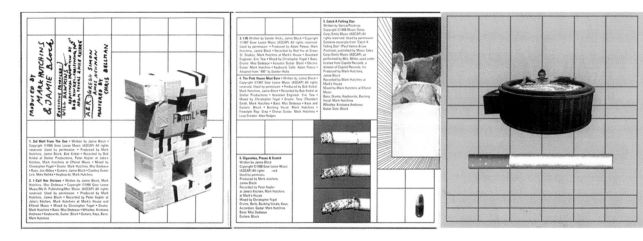

Figure (xlv)

Figure (xlvi)

Figure (xlvii)

Figure (xlviii)

Figure 5-13.
200-year-old Qing Dynasty archway Ningbo
Assembly Hall.

perceivable passage of time suggests a
commonality with Tadao Ando. However,
Mi Qiu's designs are more overtly emotional.
His use of the figure and particularly his
repetitive use of wings—a form that has
become his signature—speak to a profound
connection to nature and spirituality. Much
like his approach to art, Mi Qiu's desire to
move between "other" worlds of existence
creates a structural narrative. These wings
are sometimes in flight—additional
anatomical appendages on faceless human-
like forms are evident in his dancing and
interlocking figurative sculptures.

  More often than not he uses a solitary
wing sculpted in stainless steel. The glass
wall and the majestic archway are suspended,
floating over a pool of standing water, which
runs the length of both structures. As noted
previously, Mi Qiu's use of form and the

contrast of materials, steel, glass, light, brick,
optical fiber, and water resonate and imply
characteristics of Tadao Ando's work. Ando's
use of materials, particularly structures
unified with floating bodies of water,
elevated those structures of concrete and
glass to an almost spiritual visual place. This
visual reverence in architectural design and
environmental design also has a direct

Figure 5-15.
Mi Qiu adding final
touches to the winged
sculpture.

Figure 5-16.
Divine Existence fiber optic, crystal, and additional stainless steel sculptures. (See also Figure (xxix) in the color insert.)

connection to Frank Lloyd Wright by creating a world within the natural one he occupies that is in many ways a celebration of form and space. The romanticism of his visual intelligence and its application cannot be concealed regardless of the thickness of the materials used.

"Divine Existence" has five distinctive structures or elements: the archway (the focal point); the streaming wall of glass; the winged sculptures in both the foreground and background of the plaza and to the left of the entryway; and a glass structure suspended from the ceiling giving the viewer a sense of form, space, and freedom; and a water element described below. Here again he employs fiber optics as the structure—crystal-like in composition—covered with distinctive linear images. This too is suspended over a body of water that replicates the circular shape above it where the form originates—water being the fifth element. This form, with glowing light at both the top and the bottom of the structure, beckons the viewer. The crystalline shard completes the transformative space in the plaza and validates the ability of art and design to serve as a stage for the humanity and performance of Mi Qiu's other worlds, worlds that are a dialogue between himself, his fellow man, and nature.

Figure 5-17.
Divine Existence. (See also Figure (xxxii) in the color insert.)

Figure 5-18.
Suspended sculpture (five elements).

There are five important elements utilized in Chinese history, expressed vividly throughout its rich culture; water, fire, earth, iron, and wood. As can be seen in this and the following examples of Mi Qiu's work, he employs them deliberately.

The seven warehouses located in the eastern part of China's capital city Beijing (The Nanxin Granary) were historically used for the storage of grain during the Yuan, Ming, and Qing Dynasties, and were listed as an important landmark by the government in the mid-1980s. In spite of that, they remained ignored and forgotten for decades. Their location in the eastern part of the city near Pen An Road, an area targeted for development and revitalization, is critical to Beijing's plans for growth and aspirations to become a truly cosmopolitan city. Pen An Road is also noted for its palaces, which remain from the Qing Dynasty, as well as other historical places such as Mei Lang Fang, Wen Tian Xiang, and Guo Muo Ruo that will continue to be important landmarks

Figure 5-19.
Divine Existence.

in the capital city. Generally we attach the conservative and formal face of the country's government to this city, perhaps unfairly so. Beijing is teeming with culture, showcasing much of what makes China such an amazing destination for discovery and travel. From family owned and operated restaurants where customers are greeted with a resounding chant of welcome upon their arrival to a genuine farewell of thanks as they walk away with "double happiness"— stomachs full of the best cuisine the country has to offer and the pleasure of the experience itself—tables full of dishes each home to the likes of delicacies that of which most westerners have never tasted. (You haven't truly experienced Peking duck until you have consumed its sweetness in the capital city.)

Descriptive words for the Forbidden City are not as easy to come by. The complex is the city's main attraction. The terms grand, magnificent, and foreboding only scratch the surface of the experience. Endless corridors abound, some of which lead to doors the size of houses; others take you to secret gardens once seen only by the emperor. This is part of the backdrop that is modern Beijing. Here in the host city of the 2008 Olympics, Mi Qiu's vision for the renovation of the Nanxin Granaries complex would come to fruition.

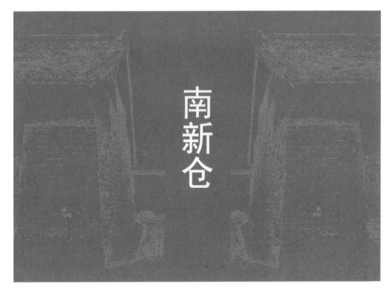

Figure 5-20.
Nan Xingcang,
Beijing project.

Five of the seven buildings are aligned on the outermost side of the complex. Mi Qiu's design calls for the existing warehouse structures to be linked in two groupings by verandas and corridors of light, steel, and glass. The dramatic use of the glass and steel corridors are reminiscent of their traditional use in Chinese history. Mi Qiu speaks often about the relationship of materials to the buildings themselves. The strength of contrasting materials is a constant design ingredient in the architectural approach to Mi Qiu's creative process. The divergent elements of past and present collide in the creation of a surreal experience for the visitor to this new complex, which will provide the public with a multitude of experiences, ranging from boutique shopping to viewing the artifacts in the museum to peaceful reflection in the sunken square and tea rooms. The square is three meters below the level of the warehouses and corridors and underscores the importance of the surrounding structures by the public from that vantage point. This was a deliberate effort to open up the space and allow the public access to a stunning view of the surrounding environment.

There is an underground corridor that leads to a sunken square. The image of the square flowing with people and commerce will conjure up past communities and the flowing system of the ancient canal transportation that was once in place here. In pursuing a balance between the past and the future Mi Qiu's design successfully creates a living dialogue between culture and history in which the public is not simply a witness but also an active participant. The history of the warehouses spans three dynasties, Yuan, Ming, and Qing. Its location was near the end of the Huang River, which placed it on the

Figure 5-21.
Nan Xingcang plan
drawing.

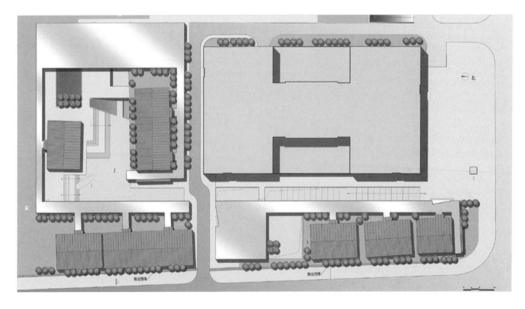

Figure 5-22.
Nan Xingcang
warehouse exterior.

main line of commerce to the city. These supplies came up-river from southern China to Beijing, formerly called Da Du, to supply the royals and its citizens. In 1260 Da Du was the center of the country, and it was also the political center. This distinction remains today, as modern Beijing is the heart of the body politic of China. As a place of commerce it was a key location as many locals owned and operated their own businesses here—making it the most progressive city in China. Mi Qiu's layout includes spaces designed for restaurants, tearooms, and gift shops that echo the

Figure 5-23.
Nan Xingcang rendering of completed site. (See also Figure (xxx) in the color insert.)

Figure 5-24.
Nan Xingcang rendering
of completed courtyard
and warehouse structure.
(See also Figure (xxxi) in
the color insert.)

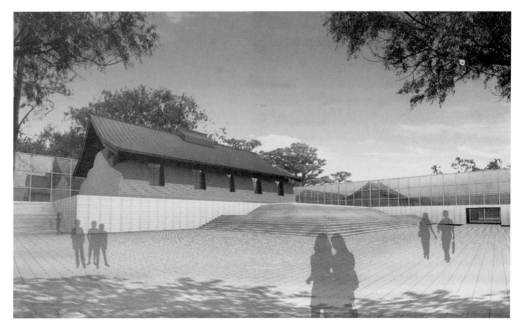

commerce and activity of that proud past. It
is no coincidence that the construction
activity in and around Beijing has increased
now that the capital city will be the host of
the 2008 Olympics.

The warehouses themselves can be seen
throughout the complex as visitors traverse
the glass and steel corridors. Mi Qiu's design
essentially turns the warehouse granaries
into installations viewed from every point
throughout the glass-enclosed corridors and
verandas. Thus, viewers can experience the

unique brickwork and rooflines distinctive
for their respective periods—suggesting that
one is witnessing something precious and
important. The sunken square is a
transcendent open space that gives the
complex the visual respite it needs. Unlike
Shanghai, Beijing's cityscape is more laid
back and spread out—the neon is toned
down and the buildings generally are not as
tall. This would seem appropriate for the
birthplace of modern China—the new must
coexist with the old. Tiananmen Gate is just
one of the many doors that are open to the
future prosperity of this city and the people
who live, work, and play here.

One of the key challenges of designing new
structures that unite the warehouses in this
complex is the development of structures
that work not only with the surrounding
landscape but also enhance the existing office
building, which at a height of fifteen stories,
anchors the complex significantly if not
overwhelmingly. This building whose
construction predates the current
development could have been an albatross
for the designer. Rather than use materials
and designs that would have highlighted the

Figure 5-25.
Nan Xingcang rendering
of warehouse interior.

Figure 5-26.
Nan Xingcang rendering of restaurant interior.

building in a negative way, Mi Qiu's dramatic use of light, steel, and glass echo the modular look of the office building, which will be the home of the Olympic Committee in 2008. The glass panels of the verandas will be tinted to match the contrasting elements in the office buildings overlooking them. These glass and steel corridors echo the past in a more historically profound way by suggesting the crucial

Figure 5-27.
Nan Xingcang additional interiors.

arteries of China's many rivers, such as the Grand Canal. The Grand Canal begins in Beijing and is the longest and oldest man-made waterway in the world and links the Huaihai, the Yellow, the Yangtze, and the Qiantang rivers—another critical link to China's economic past and future. These links to the country and its culture are deliberate connections which inculcate the land and its people in the very structures they will occupy. Five elements, five waterways—the connection and relationship of one time period to the next, this is the palate of Mi Qiu's design process.

Ground has been broken for the new complex at the Nanxin warehouses and granaries located in the eastern section of the city. The completion date is set for 2006, two years ahead of the Olympic Games. As Beijing and other cities rich with history and promise continue to grow and to develop public spaces, the need for creative and culturally responsible planning will remain a constant. There are many indications that the government and the

industries possessing the power and resources to continue fueling this growth are doing so in a responsible and thoughtful manner; however, key to that success is their ability to engage China's talented visual artists and designers.

A Celebration of the Land and its People—Yesterday, Today, and Tomorrow

It is clear that Mi Qiu holds an enviable position as one of China's most successful artists. As a member of and arguably a leader in China's first generation of artists to fully engage western contemporary art, Mi Qiu's artistic journey is thoroughly infused and informed by that experience. The mind-set of many modern artists is both the collaborative experience, and the ability to work alone in reaching one's goals. This is a key focus of Mi Qiu's creative process, which is always informed by his knowledge and experience as an artist. He speaks often of the importance of his work possessing a certain "feeling," one that grasps the importance of public spaces in a restorative way. Projects are usually commissioned after requests for proposals have been reviewed in the same way as they are in the United States. However, in China, if after all the proposals have been made and there seem to be no acceptable design proposals, then the business or the government has the option of selecting another designer for the project. More often than not Mi Qiu has been the recipient of that commission. He begins only after he has reviewed the complexities of the job and the challenges that are inherent in a project of this magnitude.

He continues to prosper utilizing his creative talents in a very broad spectrum of environmentally related design projects that include the development of new communities such as The Blue Mountain

Figure 5-28.
Nan Xingcang glass enclosed corridors.

Figure 5-29.
Mi Qiu promoting the
visual arts in the
community.

Project, where he teamed with one of China's most successful young developers, Ding Chang Feng of Vanke Co., Ltd. in creating a luxurious community for the new wave of upwardly mobile Chinese professionals. The project included an art museum, residences, and a sculpture park and is based on the notion of a full circle environment (encompassing living spaces that include areas for in-home offices and studios).

The climate remains good for environmental design and real estate development in China. With creative minds and leadership like Mi Qiu's, along with China's new wave of young and progressive developers such as Ding Chang Feng, thoughtful design and development of urban and public spaces reflective of the quality of life that can now be attained in China are being created. These individuals know from where they have come, assuring that the past is respected without short-changing the future at the expense of the present.

This linking of past, present, and future is the essence of Mi Qiu's creative process and can be seen throughout China as evidenced through the many projects that bare his trademark signature. His recent selection for the 10th Annual Architecture Exhibition at La Biennale Di Venezia for his breakthrough redevelopment project of the historically significant Watertown of Tongli serves as further validation of his importance on an international scale as well. The instillation depicting the essence of the Tongli project garnered an overwhelmingly positive response from visitors to the site, both the public and the media. Utilizing essential elements in Chinese culture along with multimedia, the exhibition design further established Mi Qiu as a very relevant designer in relationship to the importance of the built environment in major urban settings throughout the world. This project creates a new city from the ancient foundation of the existing location, which in many respects is the Venice of the east as its waterways mirror that of the famous Italian oasis. New and existing residences and public venues are being developed to enhance the city as well as amplify the location as an important tourist site in addition to being an amazing city to live in. The project is being done in collaboration with Italian architect Francesco Moreano and is in keeping with the mind-set that will keep Mi Qiu an active leader and participant in the continued growth of Chinese society.

Chapter 5 Summary and Exercises: Environmental Design and Artistic Expression for the Built environment

Mi Qiu's embrace of China's past, present, and future; its public spaces and structures in the built environment; and his understanding the people who traverse them are elemental strengths in his creative process and professional practice. His commitment to cultural conservation, history, and the public who experience his structures and environments today and tomorrow define his sincerity as a human being and are validated by his exceptional vision as an artist and designer.

1. Designer's Keys to Success

Mi Qiu's ability to create contrasting forms, evoking a perceivable passage through time, give his structures (both architectural and sculptural) the ability to inform and engage the public who uses them, thus becoming intuitively interactive spaces. In their most rudimentary interpretation they are classic examples of ABA form—repetition and contrast in composition and architectural structures in public spaces. The effect of his environmental designs on the user is manifest as structural narrative. Mi Qiu successfully employs artistic expression in the built environment as a distinct means of communication of historical and cultural awareness.

2. Historical Passages: Creating a Divine Existence

In the development of the Divine Existence Plaza for the China Life Insurance Company, Ltd., Mi Qiu's purposeful employment of five distinctive elements anchored to the two-hundred year-old Ningbo Archway engage the public in a way that cannot be ignored. The juxtaposition of these elements in the design is a dramatic transcendent expression communicating China's past, present, and future. This communication is wrapped in contrasting historical and architectural forms of artistic expression in contemporary design for the built environment.

3. Visual Dialogue in the Built Environment

A consistent thread in Mi Qiu's creative process is the relationship of materials and structures, one unto another in his environmental and architectural designs. These divergent elements are typically employed as representations of past, present, and future worlds—successfully creating a dialogue between culture and history that the public is witness to, and which they also actively participate in.

Design Exercise

Identify an existing public space that has functioned primarily as a conduit for public access (from one venue to another), i.e., public parks and buildings, train and bus stations, etc.

- *Develop an environmental design project that effectively highlights the importance of the space and its uses while also improving the environment and the public's access to it.*
- *Incorporate historical and cultural elements that are specific to the population (past and present) in the design.*
- *Ensure that the design for and of the space is a reflection of the population that will use it as well as indicative of historical and*

cultural elements that celebrate an open dialogue of these attributes.

Professional Practice

Explore opportunities that may exist in the effective revitalization and focus of historical and cultural elements in the community of your professional practice.

- *Consider environmental design projects that will serve the public as an active*

dialogue with historical and cultural issues.

- *Ensure that these issues are socially relevant to the community that the environmental design will serve.*
- *Develop environmental design projects that are reflective of the communities they will serve, while infusing a specific sense of culture and history in the mind-set of the general public.*

Fashion Design and the Power of Visual Contrast: California Surf Wear, Victorian Chic, and Glamorous Sensuality

Patrick Robinson, Creative Director and Fashion Designer

Armani Le Collezioni for White Label, Milan, Italy | Anne Klein, New York City, Perry Ellis Women's Wear Collection | Paco Rabanne

Figure 6-1.
Patrick Robinson.

Introduction

Who knew that the enterprising and talented sweat-equity of a Baldwin Hills, high-school senior would turn surfing and making surfer trunks for his friends into the inspiration that would launch a career in fashion design and lead to the position of design director for Armani's White Label. Patrick Robinson not only directed Armani's White Label, he resuscitated a dormant women's wear line on the verge of extinction and pushed it to number one in the world. This was the prelude to even greater success heading internationally known labels like Anne Klein, Perry Ellis, and Paco Rabanne. This success puts Robinson in some very elite company. It is rare for a young designer in the competitive world of women's wear to move as rapidly as Robinson does. His creative process is specific to the labels he has headed as each demanded approaches unique to the brand.

Overview

Patrick Robinson was a military brat. Born in Memphis, Tennessee, his father was a medical doctor for the armed forces during the Korean War. He was eventually stationed in Germany, safely away from the front lines, and moving his young family across continents. While this was young Patrick's first European experience, it would not be his last. Typical of many families in the armed forces, the Robinsons would move several more times before settling down in California. This is where Patrick spent his formative years; he became a surfer and fully embraced the surfer lifestyle of Orange County. It mirrored his disposition—upbeat but laid back. Patrick's optimism served him well as it translated into an outgoing personality. Looking back, it is clear that the seeds of success had begun to take root. Admiring the Quicksilver brand of surfer clothes popular at the time, Patrick Robinson quickly came to the conclusion that not only could he design and make clothing, he could do better. In creating his own line of surf-wear he took the first step into the world of fashion design and sales. As important as the creations themselves was the fact that they were developed for a specific market.

After completing high school, Robinson was supposed to follow in his father's footsteps and apply to medical school. His parents had been observant and attentive to the development of their talented son, particularly his father, who had not been given a choice of study. He had to pursue a career in medicine. His father remembered that he had once had creative desires and wanted to pursue fine art with a keen interest in painting. Given the family pressure to become a doctor, he had ignored all thoughts of fulfilling these dreams so as not to disappoint them. With this memory firmly etched in his mind, he did not want this to be his son's fate.

Robinson had a sustained interest in fashion design. His natural aptitude and talent impressed his parents, both of whom were hardworking and extremely supportive of their five children. This support was not affected by their divorce; the Robinsons were able to work together amicably regarding the custody of their children and agreed to raise them jointly in two distinct phases. His mother had taken the first phase and was the custodial parent until the children reached high school. At that point, they all moved to their father's house after completing middle school, and his turn began. Reflecting on this period, Robinson confided that he couldn't say which phase was tougher for his parents, admitting to the challenges that he and his siblings posed for both parents.

After successfully completing his senior year of high school, and under the tutelage of his father, he went on an extended visit to New York City's Parsons School of Design to get a feel for the institute of creative learning and the town he would grow to love. The school and the city proved to be the perfect fit; that September the seventeen-year-old aspiring designer began his undergraduate studies at Parsons.

A key element to Robinson's success was his willingness to apply himself and work around the clock as he began to extend his creative reach. He had fully embraced the lessons of self-reliance from both his parents. He interned at as many fashion companies as he could while maintaining his studies. He states: "It was clear to me once I got to New York City—the heart of the fashion industry, that it was not only going to take talent and good ideas to make it—I knew that I was going to have to work extremely hard to be successful in this business—and I was and continue to be willing to do that."

An American in Paris

Taking advantage of the study abroad programs offered at Parsons, the aspiring designer took off for Europe, spending two out of the four years of his academic study working and interning at many of the famous fashion houses in Paris. He eventually landed the job that "made" him, to use the vernacular. His experiences in Europe amounted to post-graduate work in the trenches and very soul of the industry—solidifying the educational foundation that Parsons developed. He was already beginning to make a name for himself as a dedicated and up-and-coming designer. Fashion design, especially in Europe, is a small world and news of promising new designers and trendsetters travels quickly. His efforts on the renowned stage of legendary fashion icons such as Coco Chanel, Yves Saint Laurent, and Valentino did not go unnoticed by other designers. The reputation of this energetic and talented designer reached the doors of Giorgio Armani, in Milan, Italy, at a most strategic time for both fashion house and for Patrick Robinson.

Robinson had worked for the Paris-based, American designer Patrick Kelly, who had burst onto the European fashion scene like a comet during the mid-1980s. A Mississippi native, Kelly shared with Robinson the experience of growing up in the south as an African-American. Kelly's love of Josephine Baker drew him to Paris. Their backgrounds were quite different, as would be their respective futures. Patrick Kelly succumbed to AIDS at age 35 as he was just reaching the height of his career. He was the first American designer to be elected to Paris' prestigious Chambre Syndicale du Pret-a-Porter, which governs the ready-to-wear industry.

Patrick Kelly's mentorship taught Robinson significant lessons about the challenges of being an American designer on a European stage; the most important lesson was to

Figure 6-2.
Patrick Robinson for
Perry Ellis Ready-to-Wear.
(See also Figure (xxxiii)
in the color insert.)

country around the world. Robinson acknowledges that this was a huge honor for him and that despite the absence of a personal life, this was an exciting and fantastic time.

This success, at this time in his career, was indicative of the good fortune that accompanied his unique talent and a prolific work ethic. Robinson did not take these factors for granted. He was riding the crest of the wave much as he had done as a surfer; however, this time the wave was success, admiration, and respect.

The news of Robinson's success at Armani traveled quickly. He was recruited and hired by Anne Klein, New York. Robinson's arrival at the Anne Klein offices in New York validated his triumphant return to the city that had helped prepare him for the profession in which he was now a major player. He had become not only a very good designer of women's wear, but he had also become a creative thinker in a fast-paced and competitive profession. The rich history at the core of Anne Klein appealed to Patrick. Anne Klein & Co. was originally launched by its founder and namesake in 1968, quickly gaining a reputation with an innovative approach to women's sportswear. The coordinating of separates such as structured blazers, pants, and accessories in a novel and sophisticated way proved to be the anchor of the brand. A direct parallel can be drawn from the Anne Klein approach to women's wear, which was historic and ahead of its time, to that of the random eclectic approach Robinson would create after becoming design director for Perry Ellis in 2003.

The differences between the two environments could not have been more profound as Robinson quickly discovered. At Armani, the boundaries were clear and the brand umbrella was large. Conversely at Anne Klein, Patrick felt that he needed to

remain true to yourself, advice that was very close to that of Robinson's father, who had bestowed the same advice upon his son.

Armani was calling and needed a spark for its failing White Label line of women's wear. In an unprecedented move, Armani hired Robinson as design director, giving the twenty four year-old fledgling designer total control. The White Label line was an international line of clothing that was due to be closed out because of its failing performance. Robinson's response to this enormous opportunity was to become totally immersed in the reinvention and transformation of the line. During this time he was totally focused on the job, giving up any semblance of a social life. The only people he saw were the people he worked with. He worked seven days a week, for five straight years, sacrificing friendships, relationships, and normal life. It was all work, all the time. However, his dedication and pursuit of perfection resulted in the Armani White Label collection becoming the number one selling collection in every create and define those elements of the company, and at that time, in his own words, he was not ready. Adding to the complexity, he had to report to three bosses—all of whom had different perspectives on what the collection should be. It was also the first time that he was totally in the media spotlight. In hindsight he notes that maybe he was not ready for this added responsibility. As his plane was landing at JFK airport, his arrival and new position at Anne Klein was being reported in every newspaper in New York City. Within two months he had been in every important fashion publication in the country. This experience, coupled with the pressure of expectations and scrutiny under the industry and media's constant coverage, was too much. Admitting he did not handle the situation optimally, Robinson would learn more in those deep valleys than he had during the height of his tenure in Milan leading the White Label collection to international acclaim. In the end, he states: "It was a great and wonderful experience in spite of the bumps, bruises, and difficulties."

The company closed after two years, because he was unable to ignite interest in the line. However, his optimistic and positive approach to life comes through as he speaks about the experience. "Dealing with and accepting your accomplishments as well as your failures go hand in hand with each other. If you are fortunate, you will have plenty of both."

Leaving Anne Klein under the sobering backdrop of his failure to move the collection forward and the impossible task of pleasing a three-headed management team, he moved on to develop a collection under his own name for the first time. This experience proved to be the stumbling block that would take the designer to a place of reflection in an effort to understand who he was as a fashion designer. Rather than brood over his disappointment at Anne Klein, Patrick did

what has become a key element in his creative process as he continued to realize his vision in women's wear. He embraced the experience and used it to start up his own line of clothing. Undaunted and focused on continuing to expand his design perspective, he still had much to say about women's wear.

The Patrick Robinson line of clothing enjoyed a moderate success, selling to every major department store in America, and included opening his own specialty store. The business was initially begun out of his New York apartment before he was able to set up the appropriate office space. And although the level was far from his experience at White Label in Milan, Robinson was not deterred and he continued working under his own label until the many demands of the business forced him to close. He was confident of his business abilities—"I get it . . . the business part . . . I just can't do it. I can smell what's right and what's wrong. It's just that it is another hat that I have to wear in addition to the creative. Addressing the day-to-day management of the company, keeping the lights on, hiring staff, making payroll, and of course trying to maintain my profit margin."

These responsibilities stretched him too thin to focus effectively on what he does best—designing sophisticated women's wear for an affordable market. Robinson looked at the experience through an objective set of glasses. "Running that business was a joy. I had a terrific staff, and together we produced some wonderful designs for major retailers. Closing the doors was very difficult and emotional for me, as my staff—having grown to ten dedicated and talented people—had become so close, working very hard to be successful in this business." The true heartache for Robinson was that he could see how to succeed—it was right there in front of him. "I just could not get there."

The business lasted about five years but his last order remained unshipped; he had run out of money to operate his company.

Unemployment

Taking a brief sabbatical from the business and being unemployed for a time, Robinson interviewed with several companies, noting the irony that he was now being interviewed by people he would not have hired. This phase of his career would take him full circle and right back to center stage. In April 2003, Perry Ellis, owned by The Public Clothing Company was looking for someone to restart their women's wear collection. They would eventually turn to Robinson to take over the reins, giving him full creative control of the women's wear division. Robinson was familiar with the Perry Ellis portfolio, and admired the late designer for the icon in American and international design that he had become. Perry Ellis had worked to create beautiful and affordable clothing for both men and women.

Perry Ellis, who died at age 46, was one of the few fashion designers to be successful in developing collections for both men's and women's markets. He had moved from the retail side of the business, where from 1963 to 1967 he was a buyer for the Miller Rhodes department store. His fashion sense and design ability in the creation of sophisticated and stylish clothing that was both fresh and innovative became his trademark stamp as a leading designer in the 1970s and 1980s. The Perry Ellis Portfolio collections were sought after by young women who embraced this new classic line of distinctly American sportswear. The success of the Portfolio collection enabled the designer to confidently move into men's wear and develop the Perry Ellis men's wear collection. This collection took men's casual wear to amazing heights with his unique styling approach.

The Perry Ellis Fashion House was also instrumental in the development of new talent, as it was home to Isaac Mizrahi, where from 1982 to 1984, was a designer in training.

Another of today's outstanding designers who spent valuable time working for the Portfolio collection in the late 1980s to early 1990s was Marc Jacobs before he reached his own fame as a standout designer.

Although Robinson recognized the significance of this opportunity and accepted the challenges of creating a new line for the famed fashion house, his initial response had been to turn down the offer. Robinson had been recommended for the job by a close friend in the fashion industry. However, he quickly realized that the people representing Perry Ellis at the negotiating table were wondering whether Robinson's personality was too much for the company and might overshadow the famous brand; perhaps he was too expensive and too much of "an all about me" designer. This opinion was perceived as a lack of focus by Robinson, who had seen this before in his experiences at Anne Klein. He left the negotiations feeling that it would not be a good fit. And the deal was off until Perry Ellis convinced him to return for further discussions.

At this point Robinson emphasized that he must have full creative control if he was to take the position. "I learned that if you are going to work for someone, in order for you to have success in your capacity, they have to have a consistent and transparent strategy for the direction of the company."

The three-headed leadership at Anne Klein had left an indelible mark on Robinson. He would not venture down that path again, even at the cost of leaving the business entirely.

For the very first time in his career Robinson considered leaving the industry. "Maybe the fashion industry had seen my day," he mused "and was ready for someone else." Embracing

Figure 6-4.
Patrick Robinson for
Paco Rabanne pap ete
Collection '06. (See also
Figure (xxxviii) in the
color insert.)

Figure 6-3.
Patrick Robinson for Perry Ellis Ready-to Wear. (See also Figure (xxxv) in the color insert.)

was about to take center stage once again, under the leadership and creative direction of Robinson.

Robinson's ideas for the direction of the new Perry Ellis Collection, and working through an open and honest dialogue session, management was able to put all egos on the table and reach an agreement that was good for everyone. Robinson's explanation of what this meant was expressive and full of wisdom possessed by one who had been atop the waves, balanced on the crest of the brutal and unforgiving fashion industry—and with all eyes on him. This time around, things would be different. Again reflecting on his time at Anne Klein where he had naively thought, "the industry was the world as you're constantly concerned with its perception of you." He had learned that the industry was not the world— "clothes are about the people who buy them in the stores—so I concerned myself totally with the customer I wanted to dress, the customer that was in my head." The House of Perry Ellis

Figure 6-5.
Patrick Robinson for
Paco Rabanne pap ete
Collection '06. (See also
Figure (xxxvi) in the
color insert.)

Figure 6-6.
Patrick Robinson for
Paco Rabanne pap ete
Collection '06.

Robinson's first showing would be the spring/summer line of 2004—the creative focus would be "chic, charming, and

Figure 6-7.
Patrick Robinson for
Paco Rabanne pap ete
Collection '06.

Figure 6-8.
Patrick Robinson for Paco Rabanne
pap ete Collection 06.

optimistic." Robinson's creative process begins with what he calls his "vision" for the line. He had the ability to see the beginning and the ending of each "story" before designing the collection actually begins. In describing the nature of the industry and how he prepares for each show, he had chosen to categorize them as stories. Choosing the words chic, charming, and optimistic he stated the message that would drive the brand in the marketplace. In his vision, every item, each piece of fabric, ribbon, and accessory must underscore the ideals of chic, charming, and optimistic; it had to be on message to be successful.

In developing his vision for the Perry Ellis Collection and his first season of shows, Robinson knew that the heart and soul of the collection would not be high-end expensive designer clothing. He arrived at a starting point that called for fashion at a great price,

Figure 6-9.
Patrick Robinson for Paco Rabanne
pap ete Collection '06.

Figure 6-10.
Patrick Robinson for
Paco Rabanne pap ete
Collection '06.

and he reinforced this idea with marketing. In his review of the recent history of the collection, this was a logical direction for the brand. Robinson began by building the appropriate vocabulary that described his vision. His catch phrase of chic, charming, and optimistic was the starting point, and the endpoint, of the process.

As Robinson engaged his design staff in the task of creating a new line of clothing he was up-beat and encouraging. He stated: "I never flat out tell my staff that what they've created for me is a bad idea—unless it's a bad idea. Mostly what I do is redirect them, steering them back to the original objectives of the line, keeping them on message."

Robinson sees his position as creative director as synonymous with that of a leader of a jazz ensemble. His job is artfully coordinating the individual notes into the

whole vision—his vision—while leaving room for improvisation by the talented staff of designers who are critical in bringing that

Figure 6-11.
Patrick Robinson for
Paco Rabanne pap ete
Collection '06.

distinctive and informed vision to the runway—and most importantly, to the closets of the customers.

Fashion Week, New York

Robinson knew that beginning his tenure with a full-fledged line of clothes during New York's Fashion Week would need to be strategic and well planned. He was confident that he could be a trendsetter; however, the impact of this approach was not only the excellence of the collection, but the timing of the show. Robinson wanted to start the week; his would be the first women's wear line to be shown. He secured the first time slot for the event, and it happened to be on a day with men's wear. The Perry Ellis Collection would be the only women's wear collection shown that day. Given these events, Robinson expected a small turnout for the show. The turnout for the show was huge and the response was overwhelming—the anticipation of the new Perry Ellis Collection by Robinson was high,

Figure 6-12.
Patrick Robinson for Paco Rabanne pap ete Collection '06.

Figure 6-13.
Patrick Robinson for Paco Rabanne pap ete Collection '06. (See also Figure (xxxiv) in the color insert.)

and the press in particular was enthusiastic about the collection and about what had been accomplished in such a short time.

The timing and placement of the show was a stroke of brilliance. Reverence for the Perry Ellis Brand remained. The tent was full and people were eager to see whether Robinson's creative engine was running on empty or on a full tank. The first show, and subsequent collections, were sublime creations and received critical acclaim by the industry, the press, and the many women who would wear them. The clothing was full of wonderful details and consistently showcased Robinson's ability to explore contrast in fabric, texture, and line. His color palette seemed to push the levels of what was rich, but still feminine. Lilac, pistachio, lavender, and magnolia were the colors of the day and saturated the bow blouses, below the knee

Figure 6-14.
Patrick Robinson for Paco Rabanne
pap ete Collection '06.

shorts, and flouncing satin tank tops. Khakis with an evening jacket and tee shirt, a rain jacket with a double collar in pink, a printed skirt and sweater with beading, underscored the wonderful detail that Robinson brought to the collection.

Most of the fashion editors in the industry are women, and their admiration and desire to buy the clothes for themselves was apparent in their positive reviews. The September 2004 show, which would be his last for Perry Ellis, followed the success of the previous three with collections that were full-blown and just as amazing as his first with a look right out of the Victorian era. Robinson once again created a provocative vocabulary that drove the creation of the collection while reinforcing the Perry Ellis Brand. The words he had written on the board in the studio of his

Figure 6-15.
Patrick Robinson for Paco Rabanne
pap ete Collection '06.

Figure 6-16.
Patrick Robinson for
Paco Rabanne pap ete
Collection '06. (See also
Figure (xxxvii) in the
color insert.)

staff . . . "Random Eclectic." The mixing of ideas from distinctly different eras or styles was the soul of the collection. "The collection is very much about today—like creating a seersucker skirt that has a Victorian feel to it. I like mixing and matching elements from different times. Like the Victorian-styled chair I recently purchased that sits next to a Phillip Stark lamp."

With "Random Eclectic" the design staff responded with texture, pattern, and sophistication in bringing Robinson's vision to the runway. "I view each show opportunity as a movie to be plotted and staged—a story to be told. This is why for me the show is not about merchandizing entirely. We will sell perhaps twenty percent of it, but the other twenty percent is about capturing the idea behind the line." Again, Robinson is telling a story from beginning to end, which resonates through the details of the clothing. "I get to teach myself, and learn something I didn't know before—every six months we start fresh."

That show was perhaps his strongest for the Perry Ellis House. During preparation for the show and in the development of the theme that would be a dominant element in his design for the collection, Robinson buried himself in anything and everything he could get his hands on from the Victorian era. From movies to biographies of Winston Churchill, he used the Victorian motif as the core of the collection, marrying tiered dresses with wide bow sashes, embroidered vests, and capri pants. He stated that, "after thoroughly immersing myself in the subject I go about the task of incorporating my very own essence and somewhat contrived aesthetic—creating a cleaner, more modern feel in the clothing. This is my creative process." That this would be his last show for the brand was

ironic given the reception for the collection and the anticipated success of sales.

From a design standpoint the designers who may have influenced him included Anne Klein. Her ability to coordinate elements that traditionally were not put together is comparable to Robinson's approach to women's wear. Robinson's collections for Perry Ellis were smart, and to use his description the "prettiest flowers." In reviewing the Perry Ellis approach to women's wear and fashion design, he would have been more than pleased with Robinson's visualization and unique approach to moving the brand forward. This is how he discusses his design ability. "This is what I do, and have always been able to achieve. That is my strength as a fashion designer. I know how to make the prettiest flowers in the garden." He makes this statement with the confidence of a man who is not a stranger to success. It has enough arrogance to give him the edge in a garden that is apt to have as many Venus flytraps and flesh-piercing prickly thorns as it has orchids.

Robinson resigned from Perry Ellis after New York's Fashion Week. A press release stated: "Perry Ellis canceled its women's wear line effective December 2004, after confirming the departure of Patrick Robinson, creative director. The company stated that it hoped to continue in women's wear in the future, but that Robinson's resignation had come in an untimely situation." The news took the industry by surprise. Details of the true nature of the split remained vague. The next surprise was that Robinson would become the new creative director of Paris-based Paco Rabanne, creating a ready-to-wear line for Fashion Week in Paris. He began his new position on December 15, 2004. While his address changed in little more than a New York minute, his process had not.

Seductive, Glamorous, Provocative

The Paco Rabanne woman envisaged by Robinson was secure in her body and sensual in her character. So read the Paco Rabanne press release for the autumn-winter 2005/2006 collection. While the tone was a stark departure from the "pretty baby" labeling of his elegant and feminine Perry Ellis women's wear line, his approach to the tone of the Paco Rabanne brand began with the proper vocabulary. He was the official "resurrection" designer of dormant fashion houses. The slumbering brand noted for breakthroughs that were more than three decades old has put its future in the hands of Robinson.

Over the years the brand had become more associated with its fragrance line than with women's wear. This contrasted with the explosive and experimental fabric use that had been Paco Rabanne's signature in the

Figure 6-18.
Paco Rabanne Fall/
Winter Look '06.

"psychedelic" year that was 1966. He was only 32, and he had dressed models like Twiggy in his revolutionary plastic dresses.

Figure 6-17.
Patrick Robinson for Paco Rabanne
pap ete Collection 06.

Figure 6-19.
Paco Rabanne Fall/
Winter Look '06.

Figure 6-20.
Paco Rabanne Fall/
Winter Look '06.

Paco Rabanne was hailed as a creative genius, particularly for his groundbreaking use of materials not normally associated

Figure 6-22.
Paco Rabanne Fall/Winter Look '06.

Figure 6-21.
Paco Rabanne Fall/
Winter Look '06.

with clothing such as plastic, metal, and paper. These were textures and materials more at home in architectural environments than in a woman's closet. That the dresses were unwearable was irrelevant; Paco Rabanne became a star overnight based on that ground-breaking collection. The legacy of the Paco Rabanne fashion house was built on his conceptual approach to women's fashion. Although his ideas were not very practical, his influences affected a generation. Today, retired at 71 years of age, Paco Rabanne happily remains in the background.

Robinson's debut collection for Paco Rabanne was well received although there were some mixed reviews. And while there is no visual connection to Rabanne's approach to women's wear, Robinson genuinely admires the elder designer, stating in an

Figure 6-23.
Paco Rabanne Fall/Winter Look '06.

article written by Kim Hasteiter in March 2005:

> It was Paco Rabanne's aesthetic that clicked much more with a provocative, sexier part of me, which I keep very personal. Paco's enthusiasm for technological advancements in fabric, and building sculptural dresses was amazing, however I will be taking the brand into a new direction.

Patrick was credited with successfully pulling off the show in a relatively short time. The dark and seductive collection was a mix of extravagance and sensual extremes including strapless dresses, draped and tapered and mixing satin, velvet, and lace. Delicately pleated knee-length skirts, some with yokes of lace paired with bomber jackets and trench coats validated the

Figure 6-24.
Paco Rabanne Fall/Winter Look '06.

Figure 6-25.
Paco Rabanne Fall/
Winter Look '06.

Figure 6-26.
Paco Rabanne Fall
Winter Look 06.

"seductive, glamorous, and provocative" theme of the collection.

Robinson is determined to remake the house with a sharp focus based on luxury. "There's more here than metal and plastic disks. There's the glamorous part, the sexy part. So let's make it sleek, sexy, and modern." Sally Singer of *Vogue* 2005 thought the show was brilliant. "Possessing all the kind of glacial chic that you want in Paco Rabanne."

And so Patrick begins a new story, this time back in the setting of his first great step forward as a designer. He has always thought of himself as an explorer. Now he will thoroughly explore the depth of the glamorous sexuality that is the future of the collection. He implores young designers to "learn your craft first and foremost. Invest in the time to get it right."

The brands have changed, but not the process as Patrick remains true to himself, while continuing to beguile the women of the world with his complete determination and ability to give them what they want.

Chapter 6 Summary and Exercises: *Victorian Chic and Glamorous Sensuality*

Robinson successfully rejuvenated some of the most famous fashion houses in the world by imprinting them with his distinctive approach to women's wear—clothing for the people who would wear them, not the industry.

1. Designer's Keys to Success

Robinson's ability to reinvent himself has contributed greatly to his sustained success. Each stop along the way has provided him with an opportunity to rethink his aesthetic and to push himself further.

2. Fashion and Jazz

Robinson equates his position with that of a leader of a jazz ensemble, whereby he artfully coordinates the individual notes into a cohesive whole—the manifestation of his creative vision.

3. Design Resurrection

Robinson's ability as a designer is equaled only by his understanding of branding and the woman he designs for. He has headed the fashion houses of Perry Ellis, Armani White Label, Anne Klein, and Paco Rabanne. Creative adaptability born out of research and a keen understanding of his market are strengths that keep Robinson ahead of the curve as he continues to redefine women's wear.

Design Exercise

Design a new women's wear line for Paco Rabanne called "PR Sport." The line will focus on active wear and sports and casual wear for young women.

• *Evaluate the competition; decide on positioning of the brand extension.*

- *Select the season that you would like the brand to have its debut.*
- *Design and implement the launch of the new line.*

Professional Practice

How does your approach to fashion design differ from the mainstream? Assess the effectiveness of your creative process.

- *Review current trends in the fashion industry in relationship to your line of clothing; update if applicable; evaluate target market and update if necessary.*
- *Develop selling strategy that works for both small boutiques and large department stores.*
- *Consider opening a boutique that sells your line of clothing exclusively.*

OXO: Award-Winning Product Design and the New Business Model

Sam Farber—Founder

Alex Lee—President

Introduction

Sam Farber is a Harvard-educated entrepreneur who successfully launched his own housewares company Copco in 1960. Today Sam is several companies removed from his first successful outing as master

marketer of tea kettles and general kitchenware products. That company continues to thrive and is a competitor as a manufacturer of quality products using innovative and popular product designers like New York-based Karim Rashid.

Farber, always a step ahead of the game, appeared to have all the ingredients for success in the kitchenware business as he has shown a penchant for stepping out of retirement to launch successful business enterprises. Copco had developed several products designed to accommodate people with some level of disability through innovative design concepts; however, those products were not marketed to their strengths, perhaps because of the challenges with marketing these items to the consumer. Thirty years later that opportunity would not be missed with the successful launch of OXO in 1990. The company won awards early on while redefining the business model and embracing the ideals of universal design in basic kitchen tools and gadgets. Between 1991 and 2002 the company's annual sales growth rate was 35 percent. Its product lines included kitchen tools, household cleaning tools, storage and organization products, and automotive cleaning tools. There are seven distinct brands that fall under the OXO name, with more than 500 products in the marketplace. The list of awards for design excellence is long and has been flowing steadily from 1991 through 2003.

Farber's background and success story is a family affair. Prior to founding Copco in 1960, Farber had worked for his father, the owner of the Sheffield Silver Company, for more than a decade. Sam Farber's uncle (Simon Farber) founded the successful company called Farberware. He himself had more than thirty years of experience in the kitchenware business when he took a sabbatical from retirement and created the now legendary company OXO. His credo of "never lose sight of the end user" came full circle and landed at his kitchen countertop in Provence, France during a summer respite. Farber's wife Betsy had a problem. Mrs. Farber found that the majority of the tools and gadgets she used in the tasks of preparing and cooking food were for the most part poorly designed, difficult, and uncomfortable to hold and use, as well as being unattractive. A mild case of arthritis made it difficult for her to hold and operate the tools—a reality shared by more than 20 million Americans—most of whom spend time in their kitchens preparing, cooking, and eating food.

Farber and Mrs. Farber were not the typical retired couple, sitting in creative contemplation around the countertop considering ways to improve the everyday tools that were causing their frustration. Mrs. Farber, along with being a devoted spouse, is an architect by profession. Both are devoted culinary enthusiasts, and their frustration was a result of personal experience. As the end user Farber knew exactly what he had to do, and asked himself a simple question. "Why do ordinary kitchen tools hurt your hands, with harmful scissor loops, rusty metal peelers and hard skinny handles? Why can't there be wonderfully comfortable tools that are easy to use? If you made them like that, wouldn't everybody want to have them?"

Like any good marketing strategist, these were questions to which he knew the answers. After spending several months in the French countryside in a rented house with Betsy Farber, cooking and entertaining family and friends, he decided to take on these problems in a more proactive way. He contacted his associates at Smart Design, the New York-based design firm he had worked with during his successful leadership of Copco. Smart Design's President, Davin Stowell, was excited and

up for the challenge. The task was to create kitchen tools that were effective, high-quality instruments, comfortable to hold with ease of use, dishwasher safe, attractive, and affordable. Stowell saw this as an opportunity to design items for the consumer that would really be meaningful. To this end the designs would be universal, appealing to a broad consumer base that included individuals outside of the so-called average user scope traditionally targeted by marketers. They would essentially be transgenerational products that are useful to the consumer throughout their lives. "This approach extends the lives of both the object and the user," Farber said. The newest marketing buzzword in product design and development "user-centered design" falls short in the mind of Farber. The term focuses primarily on the physical and the cognitive, whereas Farber prefers to use the human-centered design, fully considering the cognitive and the physical skills and including social, cultural, and emotional considerations. This progressive thinking and boundary pushing approach was not new for Farber. "Transgenerational design extends the life of the product and its materials by anticipating the whole life experience of the user." This thinking is manifest in all OXO products, where form and function excel.

The essence of universal design is that objects meeting these criteria do so without feelings of stigmatization by the end users. The physical strengths of these products, while infused in the design, are in fact invisible. OXO International is a leader in the manufacturing of products embracing universal design and has become an icon for this forward thinking and inclusive approach to everyday kitchen and household products created for an increasingly diverse population. This population includes people from all age groups and abilities. We will review the history here with a specific focus on the imaginative creative and marketing approach of a company that has redefined the successful business model.

Farber was in the second year of his three-year contract with the new ownership of OXO having sold the company in 1990. That agreement allowed for the smooth transition of the company's management and direction. That transition, strategically navigated by Farber himself enabled him to hand select the individual who would eventually became the company's president. Farber handpicked Alex Lee in 1994.

Overview

When Alex Lee walked off the Manhattan campus of Parsons all roads seemed to lead to a career as a product designer. He had completed the notable design program in exceptional fashion. That program, which was part of the environmental and product design department at Parsons, reinforced his goal to design kitchenware products for everyday use. He had always enjoyed cooking and was very conscious of the creative challenges and opportunities working in this area would create for him. Lee was ready to pursue his professional calling when he applied for a design opening at the Michael Graves firm (product design studio) as head of the model-making department. The successful Princeton, New Jersey-based architect was just at the beginning of his ascension into product development and design. These were the 1980s, a time associated with excess and the vestigial yearnings of the powerful yuppie market. That Lee would spend his eight years working in the studio of Graves, who was to become one of the most successful architects to cross into consumer market driven product design,

was fortuitous. He would learn what he did and did not want to do as a product designer.

Beautiful Products that People Used Everyday

Lee wanted to create beautiful products for everyday use that were actually used. "I realized that people would buy Michael's tea kettle (designed for Alessi) and sit it on a shelf instead of the stove for daily use. I wanted to make these products that were both aesthetically pleasing—within reach from a pricing standpoint and also practical—people would use them everyday."

Granted, the tea kettle spoken of here was quite popular, one of those coveted items associated with the term "yuppie porn" when describing what high-end designer product were to the specific market that could afford them at the time. During this period something else was formulating in the mind of the young designer. He realized that the design profession may not have been ready for the ideas that sat well outside of the traditional business approach and clearly differentiated Lee from his peers.

A closer look revealed that his father was a successful businessman and his sister worked on the brand side of business for Procter & Gamble. She was also a graduate of Harvard Business School. These family ties should not be discounted. It is not suggested that by association they had a direct bearing on Lee's ascension in the business, but that maybe there was some predisposition at work here—that unknowable thing that often runs through families manifesting and breeding success.

Lee was using his time wisely. Although not a true epiphany, he remembered an experience during his sister's time as a Harvard business student that changed his perspective. During one of his sister's vacations from school, she brought home a classmate who had also completed their undergraduate studies design at the Rhode Island School of Design. This was both a surprise and at the same time encouraging for Lee. His perception was that business school—particularly Harvard Business School—was not the place for designers, but combining a creative background with a distinct focus on business appealed to him. Lee's decision to pursue his undergraduate degree in design was the lesser of two evils in that he originally wanted to pursue a career in music. While a career as a designer does not historically fall into the category of preferred occupations by parents sending their sons and daughters off to college, it was a much easier sell for Lee to make to his parents than that of a musician playing nightclubs and traveling from town to town not knowing where the next gig would come from.

Much like a bull in a china shop, Alex considered himself to be the "modernist" within the design studio of Michael Graves Design. While this was clearly a crucial experience for him, at the time he felt out of place. He speaks of Graves with respect, but notes that they often had spirited and interesting debates. Lee was on the less decorative and more functional side of design—a true modernist at heart with the idea of creating products for the mass market. A review of Graves, the architect, suggests that his design approach leant itself more to the use of heavily ornamented constructed buildings, a post-modern nod wrapped in classical structures-an influence from his early days during time spent at the American Academy in Rome. More recently, however, products that he has designed for the large chain store, Target, take on a more modernist approach like his beautifully realized paper shredder that shares the graceful contoured lines reminiscent of Karim Rashid's wastepaper basket for Umbra, artfully named the "Garbo-Can."

The integration of the shredder at the top of the can is beautifully realized and sublimely functional employing the use of two buttons for total operation, one providing three functions (off, auto, and reverse) as well as being equipped with a pencil sharpener located directly below the shredding slot. The other button allows for clearing the shredding path.

Lee was determined that he would not languish in any design studio "bumping his head" on the ceiling of product design. He decided to go to graduate school, where he would successfully complete the business program at Harvard Business School. During his studies it was clear that he was concerned with "the widget"—an expression used in business school to describe the product. He was equally concerned with the business model, which was the only focus of his peers. He had also decided that design would be his means to an end. As a skilled and talented designer, his goals and ambitions far exceeded any experience he had had or would ever have working solely on the creative side of the business. Lee needed to be in a situation that would allow him opportunities without boundaries or ceilings.

Black & Decker and the Tools of the Trade

While attending Harvard Business School, Lee carefully began the task of researching companies that he associated with relevant product design and development. During that time the one company that consistently presented consumers with superior products was Black & Decker, the company known for its award-winning consumer power tools—this was his company of choice.

Black & Decker created the first power drill in 1915 and developed equipment that was used during the first lunar mission. It has gone on to develop products that include outdoor and indoor home cleaning tools, although the consumer power tools division remained the heart of the company. Lee had done his homework and was genuinely impressed with the product lines that the company consistently produced, concluding that he could thrive in such an environment. Black & Decker had carefully built a reputation for quality products that are well-designed for the marketplace. Lee got a job at the company, and focused on marketing and new product development. It did not take him long to realize that his time at the company would be short-lived. His perception of the organization as progressive and forward thinking from a design and product development perspective was skewed.

Lee's time at Black & Decker yielded discouraging results, but moved him forward to pursue other opportunities, including starting his own company. The defining moment came when the marketing group of Black & Decker began the creative review process, which involved an open dialogue with the designers about the products they were producing. Lee found that if he merely questioned the direction of an idea for a design, or if he made a suggestion about these products to the creative staff, the designers would quickly alter their submitted designs without discussion or further query just to appease him. This was not the business model for design and development he had in mind. Lee's view, similar to many progressive thinkers in a creative environment, is that there must be dialogue coupled with a sincere opportunity for collaboration and query in the creative process to allow for rich discovery. This ideal is often not realized in traditional settings where the marketing and management groups hold the reins tightly on their respective creative departments.

Organizations that understand the importance of the creative process and its positive effect on the bottom line, strategically link marketing and design to allow for the critical engagement and open collaboration at the inception of the project. This approach to design—assuring the active participation of everyone in developing thoughtfully designed products for consumers—is the first step.

Today, design is the final battleground in the quest for consumer loyalty—an ideal shared by many on both the design and business side. "Today's products are better, and the distance between the products and their competitors' products is small. More companies are using design to differentiate their products, not just the aesthetics of the "widget," but the total customer experience." Almost forty years ago at the International Design Congress "Profit By Design" (circa 1966), which took place in London, A. Wedgwood Benn, MP the Minister of Technology at the time, discussed the issue of design as a key component to profitability in the marketplace. "The designer must know why it is needed and exactly what it is expected to do. Part of his or her job—like that of management—is critical analysis. Good design thus has a direct bearing both on profit and productivity."

Wedgwood Benn addressed the importance of design to the success of the product in the marketplace. This underscores Farber's approach as well, that of human-entered design, which fully considers the cognitive, physical, social, cultural, and emotional characteristics of the consumer experience with the product.

Lee was ready for the next step. Black & Decker was one stop on the way to that opportunity—bringing him closer to his business model, which was an environment where the widget and the consumer took center-stage in the creative process while directly increasing the bottom line. On a recommendation from Peter Lawrence of the Corporate Design Foundation, Lee sought out Sam Farber, who was in the second year of his phase-out as OXO's president. Having successfully grown the business and turning OXO into the little company that could, Farber's goal was to select the right person to lead the company into the future. Lee was amazed at what was the main office of the company. It was a modest space, barely enough room to accommodate Farber, Mrs. Farber, and one other staff member. The award-winning OXO products were not being created on the premises.

When Farber created the company, he engaged the talented creative firm known as Smart Design to develop and design the products he marketed under the OXO name. To keep costs down at start-up, the two parties arrived at a unique working arrangement that set a successful precedent. This became a business model highlighted by Harvard Business School. Smart Design agreed to work on a royalty basis, forgoing the creative fee up-front for the long-term payoff that the product would garner. The royalty agreement was a unique way of doing business as design firms establish creative budgets that were structured around the cost of initial creative concepts at the front end. This formula was a strategic move, along with selecting and working with talented and forward thinking design firms like Smart Design and consulting relevant individuals like Patricia Moore, a noted gerontologist whose understanding of users with special needs helped OXO become a role model in the development of products that embraced universal design ideals.

Farber would hire Lee, and as his track record indicates, Lee knew a good thing when he saw it. He had found the perfect environment for his forward thinking management and business style, buttressed by his love and keen understanding of the ability of widgets.

From Tea Kettles to Measuring Cups—The Delicate Balance of Form, Function, and Durability

OXO consistently manufactures consumer products for the marketplace that are functional, easy to use, affordable, and beautifully realized. Key to this success in product development and design is the company's ability to identify and form long-term relationships with the best design firms in the world. Their seminal relationship with Smart Design has been spoken of here and represents a special historical connection, but OXO works with several multidisciplinary design firms. TODA, located in New York City, is a design firm with expertise in architecture, graphic design, and product design. Described in their own words, they seek "the possibilities that lie in between."

When the various areas of design practice in which TODA excels are combined, the opportunity for discovery is profound. These

Figure 7-3.
Original grater 1991.

unexpected occurrences lead to the development of arresting ideas that have placed the company on the highest level on an international scale. Featured in many design journals, such as *ID, Graphis,* and *Communication Arts,* the firm is also a recipient of the coveted Red Dot Award for Design Excellence in Germany.

When Lee called on TODA with the goal of expanding OXO's stovetop kettle offerings, the total rethinking of the kettle was their response. Given the popularity of stovetop kettles over the years and the abundant variety from which to choose, one could conclude that there was scarcely room for new frontiers. TODA started from the beginning and examined everything that had come before. At that point the entire focus was centered on the basic function of the kettle—how could it work best in the kitchen

Figure 7-4.
Swivel peeler.

environment? The initial sketches that were faxed to Lee demonstrated the operational concept of the kettle, which was exclusively housed in the handle. The design was in development for more than 24 months; it was called the Uplift and was pure, functional beauty.

TODA's design effectively moved the stovetop kettle to another level, merging form and function in a package that was as visually compelling as it was easy to use. The base of the kettle bears the familiar shape and stovetop stance of standard whistle-free models typical of some of the earlier kettles created by Copco and others. The genius of the Uplift design is at the handle—the only point of contact needed by the user to lift and pour from the kettle. Lifting the full kettle the user hears the blunt sound of the spout lid opening, allowing for the safe delivery of hot water. Bringing the kettle to rest, the user simply pushes the handle down gently and the lid closes with the crisp sound of the hard plastic lid engaging the brushed aluminum spout. TODA's concept for the

Figure 7-5.
Uplift tea kettle. (See also Figure (xxxix) in the color insert.)

Figure 7-8.
Jar opener.

Figure 7-6.
Whisk.

Figure 7-7.
Soft handle can opener.

Uplift kettle was safe, easy to use, and competitively priced in the housewares market. The kettle is marketed under the "Good Grips" line, which is one of seven product lines created by OXO.

The Design of Measured Intelligence

OXO's unique business structure enables the company to work with a diverse group of design firms that have been carefully selected for the long-term relationship that is key to OXO's culture of creative partnerships. It is a major component of their creative process. The New York City-based Smart Design firm has a special connection in that they were the original firm to be engaged by OXO—winning awards right out of the starting gate in the early 1990s.

Since then Smart Design and OXO have created and developed many products that have been successfully marketed under the OXO brand name. They have further extended the line with the creation of the OXO "I" series. However, it was a Cincinnati, Ohio-based toy developer, Bang-Zoom that realized the inherent flaws of the traditional measuring cup. They recognized that people were never quite sure that the cup they poured was in fact a cup. Using a traditional measuring cup, once filled with liquid to the desired level, you were never confident of its accuracy. I can attest to this as I would constantly adjust the amount because I had to stoop down to see if the measurement was accurate or, worse yet, hold the cup up to eye level. This fueled the creative minds at Bang-Zoom and moved them to the development of an angled measuring cup that enabled the user to see the exact amount from the top down. The original prototype created by Bang-Zoom, was made from an opaque plastic. Wanting to refine the patented idea, OXO called on Smart Design to take the concept to the next level in preparation for production.

Forging strategic relationships is a key component of any creative process, and Bang-Zoom, Smart Design, and OXO utilized their strengths to develop a new and improved measuring cup.

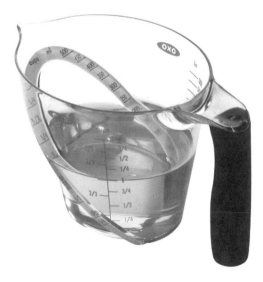

Figure 7-9.
Angled measuring cup.

A crucial selling point of the new cup would be that the primary benefit of the angled cup was clearly apparent to the user. The design team's exploration of manufacturing methods led them to use an angled ramp with the measurement markings on the inside of the cup. The clear plastic strip that runs from the bottom of the cup to the top takes the shape of an extended oval with measurements printed on the surface of the clear side with a solid white background. This gives the red type the needed contrast.

OXO discovered during this process that many traditional cups on the market were inaccurate. OXO involved the New York City Department of Consumer Affairs Testing Lab to ensure that the OXO angled measuring cups were accurate. On the recommendation of the lab, OXO engineers spent long hours in the company's testing labs utilizing scientific graduated cylinders and precision height gauges, ensuring that each marked measurement was precise.

The overall appearance of the measuring cup does not veer too far from its predecessor; it is made of clear plastic with the traditional use of the red type on each side of the cup indicating the exact measurements. At that point the form makes a distinct departure in that its overall shape is more of an oval, with a stepped indentation that is the measurement ring on the inside of the cup. There is also the trademark Good Grips handle. The large soft black cover of the handle stands out against the crystal clear plastic of the cup and complements the accentuated red sans serif type of the numeric measurements in cups (ounces and milliliters), which print on a white background strip on the interior of the cup. The design is functional, attractive on the countertop, and priced affordably—less than ten dollars for the four-cup size. The new design measured up to the critics as well as winning IDSA's Industrial Design Excellence, and the Good Housekeeping "Good Buy Award," thus validating its smart price point and its design excellence.

The design of the Uplift kettle and the angled measuring cup exemplify OXO's commitment to product excellence, which is based on form and function. OXO's ability to consistently create products for the marketplace using their unique business model and approach validates their position as a leader in making award winning products. The relationships that began with Farber and design firms like Smart Design and TODA continue to prosper under the leadership and foresight of Lee.

Lee states: "We look at existing everyday products in the marketplace and determine how they can be improved, insistently asking the question, how can this product be better at the very thing it was originally designed to do?"

The OXO products in the marketplace each have their own compelling story, based squarely on the form and function of each. The Uplift kettle and the angled cup represent only a small fraction of the success story of a company driven by good, thoughtful, and visually compelling design.

The Proliferation of Design

With the explosion of design as this new thing that we all now pay attention to and the new products borne out of this revolution in the marketplace, the consumers benefit by more choices and better designs. Perhaps life has once again imitated art as in George Perec's 1965 book "*Les Choses*," where the novelist depicts a young married couple who have an insatiable need for the best and most beautiful objects they can possess. This behavior mirrors the public's interest in products, which began after World War II, and has not faltered.

The impact that well-known designers have made on product design has resulted in the creation of consumables that are beautiful if not always functional. Consumers are willing to pay for these advancements in the aesthetic par excellence and this trend is likely to continue. Leading this surge in the proliferation of product design is Philippe Stark, Michael Graves, Karim Rashid, and others less known for industrial design and better known for their sense of style.

Society's interest and desire for beautiful items has grown steadily since industrial design came into prominence as a distinct discipline, dating back to the 1930s and through the 1960s with America's design icons Norman Bel Geddes, Henry Dreyfuss, and Raymond Lowey, the man attributed with starting the field of industrial design. The Czech-born Designer Ladislav Sutner also successfully pursued the design of everyday household products, where he

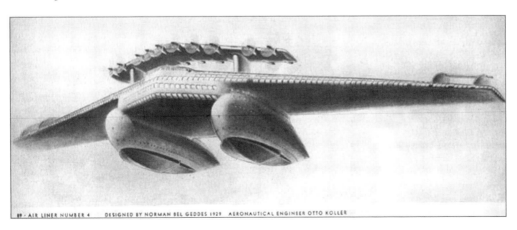

Figure 7-10.
Norman Bel Geddes airliner 1929.

Figure 7-11.
Norman Bel Geddes
airliner #4.

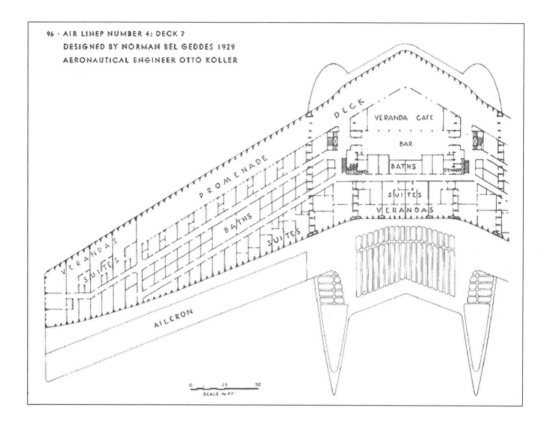

focused on the social function of his
beautifully conceived designs as well as the
honesty and usefulness of the materials used
to produce these products. Books have been
written about each of these historic figures of
modern design respectively, and we can draw
a direct parallel from Henry Dreyfuss and his
forward-thinking approach to design that
consistently focused on the user and the
practicality of the product—to OXO.

A modernist in his approach to design,
Dreyfuss was a pioneer in his use of
anthropometrics, the study of human
dimensions and capabilities. He introduced
it in the 1950s as a key element in his
business practice. His professionalism and
collaborative working style with clients and
product engineers resulted in his creation of

Figure 7-12.
Henry Dreyfuss'
streamline design of
the 20th Century
Limited 1938.

Figure 7-13.
1945 Western Electric model 302
with thermoplastic case.

Figure 7-14.
500 Dialphone introduced 1949.

superior products designed for a broad range of consumers. His Trimline phone design for Bell Telephone Laboratories is the very definition of simplicity and uniformity—modern design reaching its highest level in the handset model of 1965. The Dreyfuss design of the Trimline phone is a good example of what today is called universal transgenerational design and predates that

Figure 7-15.
A 220 Trimline rotary desk phone, with innovative dial and moving thumbstop 1965.

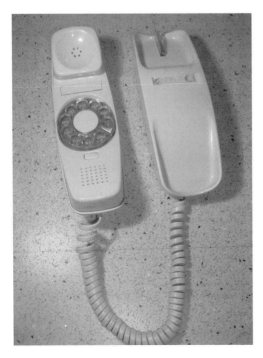

approach to product design and the user-focused culture that permeates the OXO corporate culture and creative process today. Dreyfuss is credited as the forerunner and originator of this process in the manufacture, design, and production of consumer products. The Dreyfuss sensitivity to consumer needs, and in the case of the Trimline his goal of designing a telephone that was transgenerational nearly forty years before the term was used, influenced future Bell Labs products as well as other products. These breakthrough ideals from the innovative mind of Henry Dreyfuss, are echoed in the universally designed products created by OXO. The corporate culture of this small company with big and smart product design ideas is one of optimism and inclusion coupled with the drive to "improve the ordinary things we use daily and throughout our lives." Lee's words resonate with years of experience and are backed-up by Farber's foundation of determination, focus, and collaboration.

Chapter 7 Summary and Exercises: Redefining the Business Model

OXO has enjoyed unparalleled success in the design and development of basic everyday kitchen and household products. The company's success includes an amazing growth rate with annual sales increasing by as much as 35 percent. OXO products consistently win awards for design excellence.

1. Designer's Keys to Success

OXO's embrace of universal design correlates with its success in the marketplace. While form and function are apparent in all the products the company has brought to the marketplace, OXO's commitment to transgenerational design has placed those products in households and museum stores.

2. The Power and Beauty of Universal Design

The essence of universal design is in each product's longevity and strength. OXO's creation of products for use throughout the lifetime of the user is the foundation for its success. This forward-thinking approach in the design and production of kitchen and household products is crucial to the company's ongoing success.

3. Smart Design

As design is the final battleground in the quest for consumer loyalty today, a reality not lost on OXO founder Sam Farber, his groundbreaking agreement with the creative firm Smart Design became a business model at Harvard Business School. It also ensured that the design of OXO products by talented creative thinkers like Patricia Moore, TODA, and other multidisciplinary design firms would continue to create products that are functional, easy to use, affordable, and beautiful.

Design Exercise

Develop a household product for everyday use that embraces the philosophy of universal design; the product can be new or it can be an improvement of an existing item.

- *Research similar products in the marketplace assessing weaknesses and strengths.*
- *Develop a product launch that will emphasize the ease of use as well as OXO level of design excellence.*
- *Ensure that your design embraces the attributes of transgenerational design.*

Professional Practice

OXO's ability to bring products to the marketplace is based on a philosophy that first seeks out a particular consumer need that is not currently being met.

- *Analyze your creative process; how does your client make decisions about new product development?*
- *Identify and highlight your design philosophy and why that approach makes your studio the best option for potential clients.*
- *Create a promotional CD and website that exemplifies your creative approach and clearly distinguishes your studio from other competitors.*

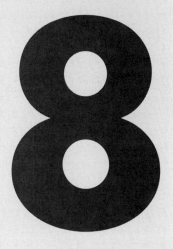

Thinking on the Outside: Graphic Design that Touches Your Heart

Stefan Sagmeister—Sagmeister Inc., New York City

Figure 8-1.
Stefan Sagmeister.

Introduction

To describe the creative mind of Stefan Sagmeister one could use the often and somewhat lame "thinking outside of the box" marketing speak of the 1990s—or borrow from today's urban culture phraseology "all that and a bag o' chips." However, this just will not do the designer justice. A more appropriate and simpler term comes to this mind: Genuine.

When Stefan Sagmeister walks into a room his tall frame and direct gaze are as compelling as the design work he has created over the past two decades, which places him among our most conceptual and thought-provoking visual thinkers. He has achieved this in the overstuffed and sometimes purely "self indulgent" world of contemporary graphic design. Stefan's physical presence alone would be intimidating were it not for the deep sincerity found in those very piercing and generous eyes—along with his genuine kindness and noble ambition to affect the world in which we live in thought provoking and relevant ways. As the title of his book states bluntly *Another Self Indulgent Design Book*—published in 2001 and written by Peter Hall—you might dismiss him as a designer purely bent on his own glorification and self-promotion. A read of the book will quell any and all thoughts of that kind. Not reading it would mean depriving yourself of a visual experience in communication both compelling and enriching, and that combines popular culture from a global perspective while encompassing and dismissing design principles in a way that is creatively redolent and engaging. To fully comprehend this experience one must submit and agree to be a willing passenger on Sagmeister's creative locomotive—with an open destination in his quest to reinvent design. This self-described goal is what makes him such a special and relevant designer and why it is a privilege to highlight that relevancy on these pages. An elected member of the prestigious "Alliance Graphique International" as one of design's most creative thinkers, the award-winning intellectual is also a sought-after lecturer. Stefan Sagmeister has been working out of his New York studio for more than a decade. The range of his visualization for the commercial market is inspiring and at the same time startling—from posters for public theaters, CD jewel case packaging, and annual reports, Sagmeister's personal approach to design continues to be arresting.

Overview

Stefan Sagmeister's formative design years started in Vienna where his first but short-lived educational pursuit was engineering. His parents were successful merchants, and the owners of a fashion retailing business in Bregenz. He was by his own admission an inattentive and uninterested student. His decision to study engineering was based solely on his natural inclination to do the opposite of what might be expected. His two older brothers had both pursued interests that would land them safely in the family retailing business. Surrounded by clean-cut pencil-aspiring engineers, Sagmeister was clearly a fish out of water, floundering for nearly three years in engineering school before transferring to a college located in Dornbirn, Austria. The college was in a thriving region of the textile industry where the local left wing quarterly publication *Alphorn* provided him with the opportunity to join the editorial team. It was in this setting that the print production of his and his colleagues' efforts were often poorly executed, nonetheless he discovered his ability in layout design and illustration. While the publication's editorial staff and business operation were loosely organized, it proved to be the perfect venue for the developing designer. Sagmeister flourished, often creating headlines by hand. Repeatedly due to severe budgetary limitations, the articles were difficult to read and the images hard to recognize. While working for the paper he became involved with local organizers of musical concerts (jazz and rock); along with the promotion and organization of these events he also had the opportunity to design the posters used to advertise these musical happenings.

Sagmeister eventually applied and gained entrance to di: Angewandt, the University of Applied Arts. His first thought was that once he became a student at this prestigious institution, his creative intellect would be challenged in ways that would broaden his outlook in visual arts and his approach to design. This was not to be. While the school's history is rich, dating back to its founding in 1867 as the School of Applied Arts of the Austrian Museum of Art and Industry, and was indeed the first of its kind on the continent, its faculty was committed to modernist thinking, heralded by the likes of Cassandre. This was not a bad thing in and

Figure 8-2.
Ronacher Music
Hall poster.

of itself; however, in the United States and elsewhere in Europe these practical tenets of design were being challenged with compelling results by April Greiman, Wolfgang Weingart, and others. His true design education would be fueled by opportunities outside of the institution's curriculum.

Sagmeister made the acquaintance of theater director Hans Gratzer, head of Vienna's modern theater. The director agreed to consider poster ideas the young designer pitched that would promote performances at the theater. Sagmeister had teamed with three fellow students to form the creative studio *Gruppe Gut*. The atmosphere was highly competitive, which turned out to be less than efficient. The short design group lasted for only eighteen months before they went their separate ways. Key to their demise was the consistent selection of designs created by one member of the team over all others. Typically each student designer would submit poster ideas to be reviewed by the creative director, who would eventually settle on a design by

Sagmeister most of the time. The situation is revealed in a quote from the director "He (Stefan) understands the problem, circles the subject, grabs it, turns it around, rips out its secret, separates it, forms it into a picture and gives back the secret."

Hans Gratzer's quote indicated a thorough understanding between the designer and a client willing to take on conceptual risks. Gratzer could see that Stefan was ahead of his peers conceptually and took full advantage of this buy giving the young designer ample room to develop unique concepts for the promotion of each performance.

This experience was critical in Sagmeister's education and that of the groups' education before its demise. It provided the best kind of professional training for its young aspirants—on the job and in the heat of the process. Typically the students would consult their professors as needed on issues of production as well as aesthetic concerns. The posters were presented to the public in an aggressive way throughout the city, displayed

to contrast with the "omnipresent melancholy of Vienna." Sagmeister's suggestive and interpretive visualization of the plays, manifest in the ability of each poster to suggest what the Schauspiehaus thought of each show, contributed to the effectiveness and visual power of each poster. This opportunity gave him visibility that he had never before experienced, while also providing the critical steps in his creative development that would propel him conceptually for what lay ahead.

As Sagmeister's creative relationship with Gratzer continued to grow, the director became very involved with the plight of the 100-year-old rococo palace, the Ronacher Music Hall, which was in danger of being demolished. Gratzer launched a campaign with the sole purpose of saving the structure and gave Sagmeister the opportunity to design the posters. The posters he created for the public theater did more than plant the

Figure 8-4.
Ronacher Music Hall poster.

Figure 8-3.
Ronacher Music Hall poster series.

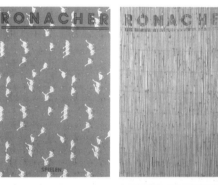

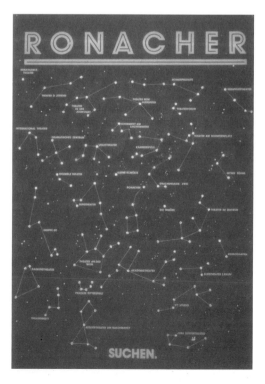

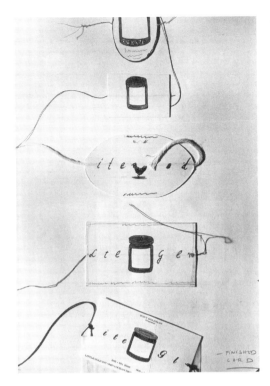

Figure 8-5.
Sketches for Tony Goldman's business cards.

seeds of salvation for the old music hall. They provided him with a rare design opportunity at a crucial time in his professional development, enabling him to create a body of work for the public arts that would have the highest of profiles. These posters focused not only on the social relevance of saving this important public venue but also provided him with creative freedom in telling the story of the music hall through the art of the poster.

Sagmeister created visuals for the posters that utilized mixed media, linked with a strong and "less is less" approach in his use of type. The word "Ronacher" appears at the top of each poster in a bold outlined sans serif font. In this way each poster is presented like a cover story, featured in a large format publication with the "Ronacher" masthead rendered on different surface treatments. There were thirty-five designs incorporated and featured

throughout Vienna emphasizing the kind of cultural excellence the theater would eventually bring to the city. The mixed-media posters, many of them thought-provoking unique designs, successfully incorporated visual elements that were innocuous, employed a sparse use of type, and married text to visual treatments ripe with a thought-provoking application of elements.

The success of the campaign accompanied with strong support from Vienna's mayor helped save the theater. Today the theater thrives and is the largest private theater in Austria.

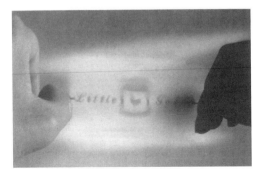

Figure 8-6.
Tony Goldman's Food & Produce.

In retrospect, this marked an important period in Sagmeister's ascension towards design relevance. As he completed his studies at Angewandt, he matured significantly and these experiences outside of the school environment were conduits for that process. He was on the path (surrounded by the asceticism of Swiss modernism during this time) to a more post-modernist approach that was beginning to light the visual fire that would spread throughout graphic design. This provided him with a platform and incubation for his conceptual approach to design.

Sagmeister was awarded a Fulbright Scholarship, which was his ticket to America and New York City. There he would study design at Pratt Institute and during that time, along with typical design school assignments, he was able to produce professional work. His scholarship provided him with a stipend that allowed him the luxury of not having to work to eat and survive. He used this limited fiscal independence strategically, and was selective as to the kind of design projects he would accept. One of the best of these projects was the trade show business card for Tony Goldman's food and produce company. Tony Goldman, a real estate tycoon responsible in part for the rejuvenation of Miami Beach needed a business card that would stand apart from the wave of business cards typically seen and forgotten at such venues. The card Sagmeister created for Goldman had images printed on both sides. When spun on its axis via an attached thread, the card spelled out the company name along with a graphic of a classic jar with a picture of a chicken on the label. It was a big hit at the trade show, and idle businessmen were seen standing around twirling the card beside the Goldman trade show booth.

You could dismiss the idea as pure gimmick without any redeeming design

Figure 8-7.
Nickelsdorf Jazz Festival poster.

attributes if you were not paying attention. The essence of Sagmeister's approach to visualization was his undying need to make you look and then look again. The classic choice of typeface, spelling the words "Little Golden," along with the rustic drawing of the jar and chicken, which complete the name of the specialty food company, suggest these elements were not arrived at by accident nor executed in a naïve way. He is naturally driven to be conceptual, but not for its own sake. His ideas are built on strong foundations and supported by critical thinking and analysis.

Stefan Sagmeister's creative process is a calculated approach derived from long periods of thinking. No wonder that one of his favorite writers is the philosopher Edward de Bono, regarded as the world's leading authority on the teaching of "thinking as a skill." Dr. de Bono is the originator of the lateral thinking concept, which employs formal techniques in an approach to creative thinking based on how the human brain functions. His concepts edify cognitive techniques that are simple, practical, and powerful methods for reaching successful and creative results. When

Figure 8-8.
Nickelsdorf Jazz Festival poster detail.

Sagmeister speaks about important influences in his approach to design and his creative process he speaks of the pivotal role Edward de Bono's thinking has played in his specific approach to design. This thought process can be seen in his work—the best of which subscribes to the strengths of simplicity, practicality, and visual power. The importance of critical thinking in Sagmeister's creative process is at the heart of each project. This process can begin with a long "walk in the park, hours spent alone in a café or park" whereby Sagmeister focuses on the blank page and begins to see the infinite possibilities.

Sagmeister returned to Austria after three years in New York City to fulfill his obligation to the government. (All men must report for military service prior to their thirtieth birthday and serve eight months of active duty.) New to military regimentation, he reported faithfully although as a conscientious objector. He was put to work in a refugee center by the military, and while this requirement had him working with refugees from Eastern Europe and North Africa, soon his creative desires lead him to graphic design opportunities in the local village. From maps to meal forms, these opportunities validated his longing for the work.

During this time he was given the opportunity to create a poster design for the Nickelsdorf Jazz Festival, which had a noted reputation not only for great performances, but for arresting poster designs in the promotions of its annual festival. After several rejections Sagmeister designed what became one of the most memorable posters created for the series. As noted earlier, his talent for creating visual communications went beyond the expected, and often challenged the viewer. His use of clever and thoughtful interactive devices, such as the golden chicken business cards underscored a design sensibility that often exceeded expectations and went beyond the standard rules. At this time he was walking both sides of the street as a designer—modern and postmodern—albeit manifested in designs that were not out of the realm of purely modernist structures altogether. While the contradictory nature of the previous statement would seem to evade the point, it is this very juxtaposition that is the essence of what makes him so salutary. It speaks to the radical insights and multiple dualities embraced by many successful designers, from Dan Friedman to Neville Brody. This creative tension during periods of graphic design proliferation (1980 through 1995) did as much to confuse as to visually enlighten the viewer.

While many intellectuals and design critics have often stated that postmodern at its best is just another hyped trend doomed to run its course, it could be argued that the opposite has occurred. Like many explosions, once the first bomb has been dropped, subsequent explosions make craters much smaller in the landscapes of our mind, having now become familiar. At its best, designs considered postmodern effectively stretch the playing field of visual communications, often exhibiting a fresh use of the vernacular that pushes ideas forward. In its worse incarnations it was just more "in your face" illegible visual noise—design for the sake of design—supported by the abandonment of content. However, when we take a closer look we find that the range of ideas and colors on the designer's palate has increased three-fold.

Describing Sagmeister with any singular label, such as postmodernism limits the definition of design sensibilities in his

Figure 8-10.
Fashion Week promotion "Litfasssaulen."

Figure 8-9.
Nickelsdorf Jazz Festival poster detail.

visual language. He *is* different, and is perhaps one of the few commercially directed visual thinkers to push the boundaries of contemporary graphic design, thereby infusing the work with images that resonate in the popular culture. These images are artfully arrived at through his own very personal identity and conceptual dexterity. His clever use of mechanical schematics along with the fearlessness of using himself, quite literally as a design element when appropriate, should not and does not dilute his power as a conceptual and practical designer. These are not one-off gimmicks set to arouse and abandon the audience or viewer before delivering the pay-off. They are visual journeys that compel the viewer to think. This brave and skillful ability to connect with people demonstrates the kind of risk-taking that is memorable and keeps Sagmeister ahead of the curve.

The poster design Sagmeister created for the jazz festival was based on a schizoid face and functioned as a lenticular—a biconvex lens with two distinctly different faces. An elastic band running the length of the poster at the top and bottom of the image (through die-cut holes) enabled it to be viewed in both its calm and confrontational incarnations when adjusted by hand. In solving the problem, or better stated, in Sagmeister's visual interpretation of the two-sided musical experience of the jazz festival's goal, he wanted to heighten the essence of that year's juxtaposition of ambient and straight-ahead jazz. The title of the 1990 poster was in fact "Nickelsdorf Confrontation '90." The three dimensional poster was a sensation and further established his ability to reinvigorate graphic design through new and interpretive forms of communication. The success of the Nickelsdorf project also highlighted his innate ability to work through his own failures in eventually arriving at an appropriate design solution. His creative process allows for this as a natural progression, although sometimes frustrating as with the initial rejections of early poster designs for the jazz festival. He meets his creative block with a playful and inventive ingenuity. His promotional campaign for Vienna's fashion week followed this path as well, proving upon its completion to be just as progressive and thought provokingly consistent with the designer's approach to visual communications. To that end, it is another wonderful example of Sagmeister's capacity to fully understand the problem, see and solve it with critical thinking, and use innovative design capabilities.

Vienna's Palais Lichtenstein was the venue for what was the equivalent of fashion week, much like that which takes place in New York where internationally known and up-and-coming designers launch their seasonal lines. Sagmeister was commissioned by the organizers of the event to create an ad campaign for the show. His initial idea was thwarted due to bad timing on the part of the media buyers working on the campaign because his plan called for the use of the *Litfasssaulen* columns found throughout the city; however, they were not available. He had decided to dress-up the bold and distinctive columns—literally. Using formal attire, he would wrap each column, partially covering the printed posters he designed for the event (each column wore a red satin-like dress or gown). Some of the columns wore deep, plunging V-necks while others sported the feather trimmed strapless style associated with award show presenters. The decision to use the advertising "*litfasssaulen* columns that the city was noted for (large columns used for informational and promotional postings) helped him turn potential disaster into an advertising opportunity. Sagmeister created a campaign that had "legs" using the agency vernacular for concepts with depth and other attributes. The ads he produced created a sensation that transcended the original media placement and that kept the

Figure 8-11.
Leo Burnet Advertising Hong Kong the controversial 4 "A's" promotion.

Figure 8-12.
4 "A's" promotional
brochure.

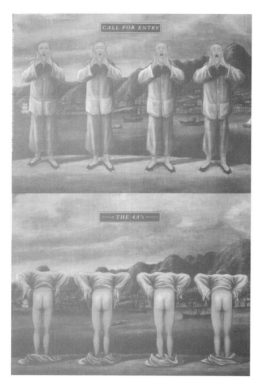

press abuzz for weeks even after the events ended.

Since it was too late to acquire the existing stationary columns for the campaign, Sagmeister came up with the practical idea of building columns specifically for the ads. This idea was truly ground-breaking and marked as newsworthy because each column was "manned" by a person who could move in and around the area most frequented by the community. The portable columns rested on a set of wheels and were navigated by the occupant. Unsuspecting pedestrians were startled when the poster and gown-clad and black-tie columns suddenly began to move while they were reading them. The students were paid for their efforts and performed well despite the difficulty in maneuvering the large mobile structures. There were sixteen columns in all and each one had a distinctive personality because of the enthusiasm and individuality that each student brought to his or her respective column. The idea was covered on television and in the newspapers

and exceeded the expectations of all participants.

Although Sagmeister's career was up and running prior to the Fulbright scholarship, which provided his passage to New York, Brooklyn's Pratt Institute, his time in the Hong Kong offices of the famous Leo Burnet Advertising agency marked a significant step in his business ascension. It also became the stepping stone that would lead him eventually back to the city that became his permanent home and creative base.

The work he created for the Burnet agency helped to define the design group within it in the transcultural environment of Hong Kong. Initially hired by the agency as its in-house typographer, Stefan Sagmeister was given the opportunity to establish the agency's standalone design group in HK. At the outset he was a somewhat reluctant hire. Traveling to Hong Kong for what was going to be a long vacation, the twenty-nine year old designer decided to test the waters. His goal was to explore the design scene in the city, which he would do by visiting studios as an applicant. Sagmeister sent for his portfolio and began applying for jobs. He had become enamored of Hong Kong with its fast pace and chaos along with the clear opportunities for growth.

Arriving at the offices of Leo Burnet to apply for a job, he was greeted cordially and offered a position after a careful viewing of his portfolio. At this point he proceeded to ask for a salary that he thought would be astronomically unrealistic, and as a result not approved. Employment was not really on his agenda at the time. This is where his naive knowledge of the business, and Leo Burnet's compensation practices unknowingly propelled him forward. The storied advertising agency has long been known for its high salaries and large yearly bonuses to its employees; this fact is as legendary as the agency's reputation as one of the world's preeminent brand architects. To Sagmeister's surprise, Leo Burnet met his salary request without hesitation. His extended vacation had morphed into a career opportunity. Initially hired as the type director for the agency, Sagmeister eventually started and directed Leo Burnet's first standalone design studio in Hong Kong. In retrospect this would seem to be an opportunity of a lifetime for him. He was just approaching thirty years of age, a milestone for a young man, and he was quite aware of his long-range goals. This phase set the groundwork for the development of his own studio one day. The work the Leo Burnet studio provided for the local businesses was nonstop, and included design opportunities within the industry as well, such as the controversial Four "A's" brochure, which made quite a stir among its participants and in the media.

Sagmeister had full autonomy, including the ability to approve or not the type of projects the design group would take. The only requirement was that the studio made a profit. His earlier work and experiences advanced him to such a powerful position at the ripe age of twenty-nine and validates his place among the field's most prolific at this

stage. The studio worked around the clock it seemed to him at the time. "We were so busy during this period that I would typically leave the studio, having worked fourteen to fifteen hours only to return and find designers camping out in the offices with toothbrushes in hand having worked around the clock."

The pace was so fast that the designer often managed as many as 125 projects at a time.

With a quick review of Sagmeister's work, it is easy to conclude that he is predisposed to producing design solutions and ideas that are over-the-top and controversial, and which shift the attention away from the event or product itself. You would be hard pressed to find any direct connection or influence of the great George Lois, architect of the "big idea" concept in effective advertising solutions. Nevertheless it is clear that Sagmeister subscribes to elements of this thinking. George Lois wanted to, and succeeded in, affecting the current (at the time) and future culture. In his own dramatic and often controversial way he has tried "changing peoples' minds." During his career he created advertising campaigns that echo today, years removed from their initial launch into the public mindset. His "I want my MTV" launch alone "rocked" our world and the music industry.

"Changing peoples minds" and "touching someone's heart," the latter ambition coming

Figure 8-13.
Block CD packaging. (See also Figure (xliii) in the color insert.)

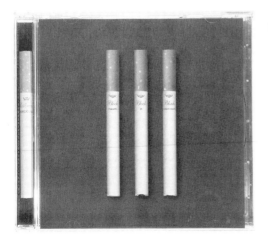

Figure 8-14.
Block CD packaging detail.

from Stefan Sagmeister, is in many ways akin to the Lois philosophy. When we squeeze the substance out of this philosophy, two salient points stand at its core: transforming our language, and affecting and therefore changing popular culture. These are indeed lofty goals that invariably place the individual bold enough to attempt them very close to the sun. Sagmeister's ability to transform the language of graphic design is well documented and continues to inform us more than a decade after the launch of the studio that bears his name. It seems more than serendipitous that he would bridge the time between Hong Kong and establishing his own studio with a short but beneficial stint at the offices of the controversial Tibor Kalman, the successful studio head of the now legendary M & Co., in New York City.

Stefan Sagmeister worked at the M & Co. offices for the better part of a year before leaving on good terms to open his own studio. The Kalman experience came during the height of that studio's most prosperous period. Recruited by Tibor Kalman, Sagmeister's application for U.S. residency status was sponsored by Kalman's offices. While he was a student at Pratt, Sagmeister, like many progressive young designers aspired to work at the legendary studio, and tried unsuccessfully to attain a position during that time. The impact of M & Co. on the culture of design during the 1980s and 1990s was significant. The company was on the front line of postmodernism's influence on the popular culture and graphic design through its clever use of visual icons and language rooted in the vernacular.

Sagmeister's return years later as a seasoned designer and former studio head gave him the opportunity to realize his unfulfilled dream from his student days. Kalman hired him as a senior designer. It was thought that Stefan would eventually lead the design business for the firm, but Kalman remained a hands-on manager preventing Sagmeister from having the true flexibility and control that he had experienced with Leo Burnet in Hong Kong. Although the opportunity was short-lived, the experience was critical in the further development of Sagmeister's approach to design. He learned firsthand from the man he respected most what would be essential in developing his own approach to the business and the culture that would define the Sagmeister studio operation. This respect and admiration for Kalman is documented eloquently in Sagmeister's book.

The most practical and biggest lesson learned by the designer from Tibor Kalman was to "think small." Keeping overhead at a manageable size while employing only essential staff allows one to keep one's hands on the pulse of the company and to maintain practical oversight of each project. Once Sagmeister opened his own studio in 1994, he never strayed from this critical approach to the management of his studio and the implementation of his business plan. Tibor Kalman's advice rang true through periods of market change while surviving the peaks and valleys of running a design business in the competitive climate of New York.

That Sagmeister shared the views of Tibor Kalman regarding the world and design's place in it is understood. This association was motivational for Sagmeister and he demonstrated it in his own business practice. Tibor Kalman's concern with the state of the world and his desire to use design as an agent for change is an area of common ground. The idea that through the power of design one has the opportunity to touch someone's heart is the culture in the Sagmeister design practice. Sagmeister's success made it possible for the firm to be very selective about the clients they chose to work with. This enviable position is not by accident, and it is

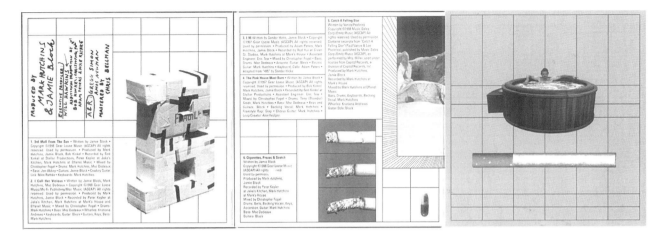

a key element in the studio's long-term business objectives.

Sagmeister's original plan was straightforward, direct, and very personal. He first and foremost wanted to design packaging for the music industry. He wanted the business to prosper while maintaining its small overhead and size, and finally, he wanted to enjoy himself, to have fun in the process of it all. His promotional announcement of the studio's opening took the idea of "fun," as personally and in your face, as an initial announcement could have ever been.

Stefan Sagmeister's creative impact on the design community came through the vehicle of music CD packaging and design. This was based on a pragmatic and strategic decision. Riding the wave of the healthy productivity of artists like Lou Reed and David Byrne of the Talking Heads, and legendary rock n' roll icons such as the Rolling Stones, Sagmeister created CD packaging that was artfully arrived at and visually motivated by his respect and understanding of the artist and the music. It was some of the early CD packaging designs for lesser-known musicians that moved the Sagmeister studio to the top of the music industry's list when it came to CD packaging and design.

The 1980s delivered the new CD packaging to consumers. The replacement of the standard LP with packaging that provided designers with a canvas a fifth of the size was a challenge deemed less than deserving of the effort. Sagmeister saw this as an opportunity and preferred the new smaller format to the larger LP format.

The designer's timing was fortuitous because productivity during this period coincided with a stable period in the music business. This changed significantly over the

Figure 8-15.
Block CD packaging detail. (See also Figure (xliv) in the color insert.)

Figure 8-16.
David Byrne "Feelings" CD cover. (See also Figure (xl) in the color insert.)

Figure 8-17.
David Byrne "Feelings"
CD cover detail. (See
also Figure (xli) in the
color insert.)

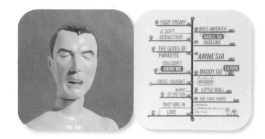

years but not before Sagmeister's creative mark was burned into the industry's visual consciousness. His packaging and design treatment for the short-lived group "H. P. Zinker" was considered by critics as truly interactive and set the Sagmeister studio squarely on the landscape of CD packaging and design. The design garnered Grammy nominations as well as inquiries for work on future releases from other labels. The design employed color filters that had to be precisely matched in the printing process for the visual effect to be successful. Utilizing two images—a man on one side depicting a calm and somewhat sedate gentlemen and on the other side a man grimacing in a fit of rage. When it was slipped out of the jewel case, the face would change. The process resulted in an increase in the print production cost that the record label was unwilling to pay. Sagmeister paid the difference out of his own pocket. This process–effect would be used in several different incarnations for different clients including the cover design for his book.

Clearly the creative stakes were high for Sagmeister in taking the unusual step mentioned above. It further validated the importance of the work and is an example to be admired. Sagmeister's ability to be visually clever is never heavy handed, holograms notwithstanding, and does not obscure the design itself. It is the sole purpose of these clever design ditties to be a means to an end—visually stunning and engaging to the viewer.

Of the many CD packages designed for the music industry by Sagmeister, the two created for David Byrne and Talking Heads are favorites. The "Feelings" CD called for the creation of the musical artist as an action figure. Collaboration is a key element to Sagmeister's creative process. Having the conceptual input of Byrne, considered by many a musical genius, was a gift to the designer, who stated: "I was willing to hear what David had to say regarding creative direction given that he is much smarter than I am, which makes him an ideal client."

Typically he would not accept a project where the design idea has been pre-approved. The concept of the action figure was pure David Byrne. Initially Sagmeister had the figure and likeness of the musician digitally rendered in a 3-D program, but this did not produce the desired effect. The image would be shown exhibiting four different emotional expressions: happy, angry, sad, and content. Sagmeister enlisted the artistic talents of Yuji Yoshimoto, a Brooklyn-based designer with a reputation in the advertising industry for the creation of giant pieces of lifelike fruit on the heads of popular professional athletes. These would be executed reproductions in 3D with authentic results. Working with clay and the photography of Tom Schierlitz, Yoshimoto molded the four heads for the cover. They were then fabricated, cast in plastic, and each doll's face was then hand painted.

Figure 8-18.
David Byrne "Feelings" CD cover. (See also Figure (xlii) in the color insert.)

Figure 8-19.
Book design for David Byrne.

The package design's concept for the CD is thoroughly realized with the creation and use of a color key "mood computer chart" essentially depicting the four emotions as well as the use of the twist-and-snap plastic device used for the text and song list that is reminiscent of a model kit. This visual device would be a printed color image included in the CD packaging; however the budget would not support full color throughout the package liner notes and booklet. Consequently the twist-and-snap plastic type was printed in black and white, still effective conceptually although less spectacular as a black and white halftone image. The color pallet and key created for the CD package was linked to each song, highlighting the emotions associated with each title.

The CD packaging design for "Feelings" is a great example of Sagmeister's attention to detail and visual delight in working thematically by using clever design elements to unify the visual experience and the interaction of the product for the end user. While this project dates back to the late 1990s, it remains fresh when viewed today.

The action figure concept was also used on a later book project, *Your Action World*, written and conceived by Byrne as a companion piece to his photo exhibition in Trieste. The book was also designed by the Sagmeister studio and, as in the case of the music CD, it allowed for great collaborations.

Sagmeister's most recent CD project for Byrne falls into the boxed-set category of packaging in name only. It is the compilation of the band's music, spanning three decades, and is a skillfully realized design utilizing a

Figure 8-20.
Talking Heads box-set CD design.

Figure 8-21.
Talking Heads box-set CD detail.

Figure 8-22.
Zumtobel Annual Report cover. (See also Figure (xlv) in the color insert.)

book format for extensive text and previously unpublished photos of the band that utterly belies production convention. The CD booklet houses all the disks in its 5¼ × 16¾ format. The set includes four disks, one of which is a DVD, while the cover features the range and collaborative paintings of Russian contemporary artists Vladimir Dubossarsky and Alexander Vinogradov depicting beautiful and naked individuals smiling, dancing, punching trees, and bleeding from strange and painful areas of the body. The illustrative cover art, with prints on the inside covers as well, is evocative of a visual nirvana—a fantasy sense of well-being, contentment, and ambiguous violence. That is, until you are confronted with the fellow who appears to be bleeding

from his genitals, with of all things, a smile on his face. A spot illustration appearing on the title page of the CD booklet shows a fallen landscape painter bleeding from a cut just below the armpit. His severed arm is in the mouth of a wolf nearby and the tool of his trade, the paintbrush, is held tightly in the hand of the artist as if ready to resume the task of painting the landscape albeit in a suspended state of detachment. "The artwork uses some of my favorite visual icons-babies, bears, severed limbs, and bare naked people—excluding the strange fellow in boxer shorts literally boxing with a tree."

And then there is the curious omission of any text on the covers of the package excluding the spine, which knocks out of a black background and sets in a sans serif font.

The gradual decline of sales in the music industry of late has been a steady and unforeseen reality for the studio. When Sagmeister opened his doors for business in 1990 the CD work represented more than two-thirds of the business and was the studio's primary focus and income-producing area. Having foresight and the need to keep moving forward and experiencing new challenges, Sagmeister redirected the focus of the studio after his sabbatical. Today the studio is broken into four distinct areas of practice: 25 percent corporate design; 25 percent socially relevant projects; 25 percent art and design projects; and 25 percent music packaging. The formula is working, as the studio continues to create stunningly effective design solutions for a significant market.

Sagmeister's strength in corporate and publication design has overshadowed the glamour of the music CD work that constitutes a stream of his design work over the years. Nevertheless his ideas in these areas are as compelling as any CD packaging he created.

Figure 8-23.
Zumtobel spread. (See also Figure (xlvii) in the color insert.)

Two projects executed by the studio come to mind here and are stunningly beautiful. The annual report designed for the European lighting company, Suitable AG, achieves rare heights in that the design of the annual report demonstrates the strength and sophistication of the lighting system by highlighting the brilliant diversity of the product at work, essentially becoming a very effective demonstration ad. The cover image is derived from a heat molded relief sculpture of an arrangement of five large pedaled flowers in a plain vase. Appearing on the cover in white, the same image is then executed on the interior pages essentially using the same shot under differing lighting conditions on each page spread. It is a clever device and signifies the power of the lighting product. With both client and designer, there was enough trust between them to execute a layout that was as conceptual and nontraditional as this annual report.

In 1999 Sagmeister decided to fulfill an earlier desire to step away from the commercial business and take a full year

Figure 8-24.
Zumtobel spread. (See also Figure (xlvi) in the color insert.)

Figure 8-25.
Zumtobel detail. (See also Figure (xlviii) in the color insert.)

consistently and critically asked the difficult and sometimes unanswerable questions: What is creative about that? And how can it be better? His creative process embraces failure as well as success, and experience has taught him that the residue of that embrace fuels tomorrow's accomplishments.

In defining Sagmeister's approach to design, which includes his distinctly personal mind-set, the best way to describe his talent and visual point of view is to begin with the word "honesty." He is an honest designer and that manifests itself in visually compelling designs that have kept him at the forefront of his profession. His conceptual thinking is consistently ahead of the curve and stands up long after the "ah ha" moment has passed. He has been true to his beliefs and to the mentorship and advice of his friend and colleague, Tibor Kalman. He has never lost sight of the "staying small" business model in his professional steering of the design studio that bears the name Sagmeister, currently staffed by Matthias Ernstberger, Stefan, and their student intern.

While one might dismiss his stated desire to "touch someone's heart with graphic design" as a hollow and romantic phrase associated with other self-aggrandizing individuals in our contemporary landscape, that would be a profound mistake. Sagmeister takes himself, and the world, very seriously. He believes "that you have to do it (graphic design) for love not money. If you do it for the love, then the money will be a by-product of that love."

Graphic design is very personal for Stefan Sagmeister, and he has shown us his scars, his passions, and most of all, his honesty.

Chapter 8 Summary and Exercises: *Rethinking the Box*

Stefan Sagmeister helped define the notion that design could be an agent for change during the 1990s while working closely with

off to concentrate on experimental projects. Closing the studio for a year and taking time away from the hectic pace of clients and deadlines proved to be the cathartic experience Stefan knew that it would be. It is another example of his creative risk-taking in search of a higher ground—essentially gambling with his company's position in the marketplace and its future.

In reviewing the rich history of this forward thinking designer, visual artist, and communicator, the thoughtful observer is not surprised by the dramatic and provocative path taken by Sagmeister. In these pursuits he has been quite consistent in his ability to look beyond his applied approach to problem-solving solutions in the practice of graphic design. His approach and outlook have

Tibor Kalman at M & Co. He also demonstrated the power of design as a vehicle of self-promotion. Since that time his New York-based studio has moved from its original creative focus of designing exclusively for the music industry to a more typical studio practice creating corporate communications for a select group of clients as well as working on socially relevant personal projects.

1. Designer's Keys to Success

Stefan Sagmeister's personal approach to design remains a strategic part of his practice today. He continues to infuse his creative process with extended periods of critical thinking, which are essential to the success of each project, stripping the page clean as he begins to see the infinite creative possibilities for a solution. He remains fearless in his willingness to create communications that exceed expectations and that challenge the viewer to pay attention and accept new ways of seeing.

2. Risk-Taking and Defining a Visual Language

Stefan Sagmeister is one of the few commercially successful creative thinkers to consistently push the boundaries of contemporary graphic design by infusing the work with images that resonate popular culture and underscore his distinctly personal handprint of conceptual dexterity and clever use of mechanical schematics and visual illusion. He compels the viewer to think as he or she embarks on this visual journey.

3. Changing Peoples' Minds by Design

Stefan Sagmeister's ability to create design solutions that are conceptually outside of the mainstream is a direct result of his

willingness to take risks. He is constantly looking beyond what is expected in his approach to graphic design, always questioning, "What is creative about that?" The results are a body of work that exemplifies his forward thinking.

Design Exercise

Select a major utility company (Con Edison, PSE&G, etc.) and begin developing a corporate brochure that re-introduces the company to its customers in an innovative and nontraditional form by focusing on their commitment to quality service and conservation.

- Research the history and corporate culture of the company.
- Analyze the concerns of customers, as well as past forms of customer communication and specific promotional materials used.
- Research alternative materials for the design and packaging of the corporate brochure that can be linked thematically to the company's products and services.

Professional Practice

Sagmeister had clear goals from the beginning for his design practice. He wanted to establish a studio that produced work that was creatively excellent and relevant, first for the music industry, and later for corporate clients and social issues.

- What were the original goals for the establishment of your design practice or job; how close to those expectations are you today?
- After analyzing your professional practice, assess the ways in which you might be able to refocus your design practice; include a critical evaluation of your design work and current list of clients.
- Commit yourself to doing work that is as relevant to you as it is to your client.

Epilogue: Design Proliferation and the Next Great Society

Pro-cess. *n*

1. a series of actions or steps taken in order to achieve a particular end
2. a natural or involuntary series of changes

Oxford American Dictionary

A casual review of the definition of the word "process" can leave one with a distinctly clinical perspective on it [the word] as a discrete means to an end. I first embraced this notion of a creative process during a graduate course taught by advertising creative extraordinaire Gary Grey of Carter Grey Advertising, head of the Toronto-based agency at the time (circa 1990). His colorful and very informed presentation on the importance of defining one's own creative process was the lynch pin of his own brilliant career, and subsequently, became the mantra of all who experienced his philosophical approach to thinking critically on a creative and very visual level. Buying into this idea would be the key element of engagement in the development of relevant creative ideas that essentially moved products and services—delivering them to the targeted group of people in the consumer marketplace. At the epicenter of that process is the individually relevant participants—the designer or art director, the client, and, of course, the people who will benefit from its use. However, before something can be moved, it must first be created. The ideal atmosphere for the development of creative design solutions must be open, inculcating discrete information directly related to the product or service and the people who will use it—and this is just the beginning. Possessing the most detailed and descriptive research information is useless without a clear vision for its application and method of delivery. The dialogue on process that is at the heart of this publication attempted to expound on the diversity of each designer's approach to her or his professional practice in a very personal

way. The playing field here was not based on quantitative information. As with many things that deal with the human touch, marketing charts and figures on behavioral patterns will only lead you so far. I will try to temper my enthusiasm for design and the critically important space it occupies in our society in an effort to be as objective as possible. (Not an easy thing to accomplish while using a gleaming PowerBook G4 with the seventeen-inch screen to prepare this manuscript, but I'll give it a go.)

It is as true as the chair you sit in while reading this book that design continues to shape our world—for better or worse—and has been doing so since our hairy ancestors crossed and bound several sticks together to fashion digging tools and weapons. The designers here are bound together as well. Their tools and weaponry are manifest in their respective creative processes. To a person, this term defined their specific methodology as regards their collective design practice—across all disciplines. Issues of innate, environmental, and societal concerns, even within a consistently defined approach, modify these processes, as does the corporate cultures they reside in. More importantly, it is what they bring as individuals to their approach in the creation of successful design solutions.

The OXO story is a positive and noble study in the personal observation of simple, but very important kitchen utensils used in the daily lives of the consumer. While Sam Farber was the product of a successful business family, it was the observations made as a consumer by him and his wife that led to the development of drastically improved everyday household products—not marketing reports and consumer testing. The critical importance of observation and seeing is not a new thing. You might think that there is a redundancy at work here with the use of both the words observation and seeing. There is a significant distinction between the two. The attentive watching of someone or something; the careful observing and recording of something that is happening, as in a natural occurrence of phenomenon is the essence of "observation" and was at work in the Farber kitchen that wonderful summer they spent vacationing in France. Seeing is another matter that, of course, is connected to the act of observing. Perception and vision, however, come into play here. These qualities and attributes are driven by differing individual factors that concern culture, environment, personality, time, and knowledge.

The multi-talented Mi Qiu's concern with the preservation of culturally significant aspects of daily life in China is a driving force in the creative process he employs, which relies on the years of observation he has banked while participating in the growth and planning of China's urban centers and the redesign of ancient cities like Tongli. Understanding the lives of the people who will be affected by these structures in the built environment, and having the vision,

creative sight, and allowing for solutions that always keep the people and their historical and cultural needs in view, will benefit them most. As with the flowing waterways of the ancient city, this is the river that runs consistently through his creative process.

While Mi Qiu heads an amazing number of successful and highly creative individuals in China today, he is not alone, nor can China afford for him to be as it continues to grow at an unprecedented and steady pace. Veteran designers like Alan Chan and Sergio Young will continue to help shape China's new society—a society that is becoming increasingly international in its make-up. Young, whose design firm Global Creative Works has designed and developed interiors for some of the top hotels and restaurants in China, Taipei, and Moscow, is based in the heart of Shanghai. The studio's focus to date has been interior design; however, they have more recently included furniture design. Respectively, they are joined by designers such as of Beijing-born Han Jiaying, whose design firm has grown steadily since its founding in Shenzhen in 1993. Shenzhen is a

Fig-01-E.
Mi Qiu working with the children of Shanghai on a community art project.

Fig-02-E.
Jiaying packaging items
for identity project for
HZ Bank.

former fishing town whose population has exploded and is China's fastest growing city; it may become its design center as well if Han Jiaying's vision comes to fruition. Han Jiaying Design now has offices in Shanghai and Beijing. The quality of the work rivals any in China or abroad. His grasp of branding and playful use of type juxtaposed to traditional Chinese calligraphy is extremely effective. Han understands the importance of these developments and is playing a key role as he pushes Shenzhen to become China's design center through his support and association with the Shenzhen Graphic Design Association. The organization hosts annual design conferences where invited speakers from around the world present their ideas on the current state of design. (Stefan Sagmeister participated in 2005 along with the founder of Europe's new design group Lava.) The growth and development of design education in that region will be a key component if it is to become the country's design center.

As China's thriving cities become the international centers that they are destined to be, the peoples' ability to traverse those centers with ease will rest directly upon the shoulders of these and other talented designers who have a clear understanding of the environment and the historical culture. They must add to this understanding a worldview—observation and seeing—as the numbers of foreign visitors will continue to increase thus, expanding the culture and vernacular in a broad and profound way.

The renowned Sara Little Turnbull understood the importance of this (seeing and observing) in the United States as far back as the 1950s when corporations first

Fig-03-E.
Han Jiaying billboard ad
for identity project for
HZ Bank.

Fig-04-E.
Tritonic designed identity project for Newark, New Jersey-based community organization.

began to take consumer issues and concerns seriously. She was the editor of *House Beautiful* after World War II, where she informed her readership of the changing culture as they began to navigate the new world that existed after the war. More recently as head of Stanford University's Process of Change, Innovation, and Design Laboratory she understands the power of design as an agent of change in society. Her belief that successful design is the offspring of culture and commerce was ahead of the great curve that in my estimation we are still navigating today. Her thinking resonates with Deborah Sussman, particularly from an anthropological perspective on design, which they both share. Sussman's approach and language in her design practice thoroughly considers the "DNA" of the people utilizing the space in the built environments that make up her canvas. The way-finding systems Sussman/Prejza designed for Hancock Park never leave the perspective of the public who traverse that space in the heart of the Los Angeles community daily.

On the community-based side of design practice the upstart multidisciplinary design firm Tritonic based in Newark, New Jersey, and headed by Andres Jeminez, Nunzio Esposito, and Luigi Tartara have thoroughly embedded their design practice in a community where change, although constant, has come at a slow pace. They have staked their design practice on the successful revival of a city

Fig-05-E.
Sergio Young Global Creative Works interior design project.

Fig-06-E.

trying to break free of its past. Tritonic takes a proactive approach by working with community-based organizations that provide services and products for the people of the city. They are effectively utilizing design as an agent facilitating positive change and growth in the community. This positioning has also led to commercial success, as for-profit businesses, such as IDT and others based in the city have requested their creative and conscientious design services where the focus is always on the people. This is the essence of Turnbull's philosophy, one that always begins with the people and their environment. What do they need, and how can their lives become better through the development of useful products designed just for them? The answers to these questions can be found all around us or as far away as the plains of Africa, where Turnbull began to realize solutions for the improved manufacturing of cooking lids while observing the skill of a cheetah's ability to grasp and hold something in its mouth while running at extremely high speeds. The suggestion here is not that you have to jump continents to necessarily find the answers to consumer-driven questions aimed at improving products; however, if your eyes are not open, you will not see the possible solutions that may be directly in front of you. In this regard, Turnbull reiterates, "The world is your oyster."

Practical consumer needs (like cooking lids that hold firmly to the hot pots they cover), will always be an evolving element in the daily lives of us all. And while design historically acknowledges change, it has the opportunity to take charge of that change on a more proactive and human level through the creation and design of a society that is more functional, accommodating, and communicative. This does not and should not come at the expense of the aesthetic or the bottom line. The human desire for all things beautiful will continue to be as important for some as the actual need and application of the thing itself. Product design of late has been called a type of pornography for the affluent—the design of products historically built for daily use like that of tea kettles, that oft-times sit on shelves to be admired as objects, not practical utilities for use in the kitchen. While there will always be product design that falls into this category, possessed by people only concerned with the beauty of the widget and not its application, this will continue to be balanced by people like Alex Lee, who sees to it that the tea kettles produced by OXO are able to comfortably and safely pour a good cup of hot water for afternoon tea, all the while enduring the rigors of the typical kitchen and providing visual presence on the countertop at an aesthetically high level. (I must admit to setting my recently purchased Uplift kettle by OXO on the countertop and admiring it for more than a day or two before actually using it.)

Considering the power of design presently, metaphorically we have seen the pitch, and knocked it deep into left field, but we have only made it to second base. Looking at things to make them better for the sake of society at large, without losing sight of the valuable bottom line would seem to be an unrealistic goal where design and business are concerned. Is there a paradox in that statement? As we have heard and seen in the past, the focus of business is on the bottom line, on profits. The easy answer is yes; however, it is suggested that this is where the challenge resides for designers and the companies that employ them in the creation of products and services for the consumer. Because smart design has shown us that customers respond to brands that clearly understand their needs, and where that understanding manifests itself in the product, the customer is sure to keep coming back. Dare I say it—the goal of good design should be that of lifetime membership by the consumer of choice. Do the right thing by the customer once, and they are sure to return. This paradox can be good if we are diligent in keeping people first in the order of things and we maintain the right balance.

Should design be socially responsible? This is what Tibor Kalman attempted to get at with his "*Colors*" publication and the generally untrusting perspective on business's use of mass media to flood the consumer market. He implored fellow designers to be more selective about the clients they choose. Stefan Sagmeister feels that as a designer he should and does consider the value of the work. The proliferation of design is moving at an eye-popping pace. The understanding that society does not necessarily need more "stuff" as George Carlin humorously cajoled during a famous monologue is an ideal that designers and businesses should embrace. We need the "stuff" we have to be better—like the present state of music. Thanks to the forward thinking of Steve Jobs, the CEO of Apple (co-founded in 1976), the iPod and iTunes is the current model of the popular forty-five records that were playing in many teenage bedrooms in the 1950s and 1960s. You did not need to buy complete albums, where often only two or three of the twelve tunes were really great. The new music technology spoken of here, exemplifies the true empowerment of the people, who benefit from the design and use of the product. This is intrinsic to the success of the iPod and the iTunes online music store.

During Patrick Robinson's successful but brief tenure at Perry Ellis, he was concerned with designing women's wear that was affordable, but beautiful to wear and look at. Focusing on the true needs of the customer was not a new thing for him as when he designed surfer-wear for his classmates in high school. The recent success of retail chains like H & M validate this thinking, because people want to look good but not at the sacrifice of their next meal or monthly rent. We can add the phenomenal success of the Target retail stores to that list as providing their customers with smart consumer based design products at a reasonable price without sacrificing quality, which is the basis of their existence. (Someone has been paying close attention to the consumer at that company as well as the power of good design.)

These critical observations and thinking skills are the keys that each of the designers featured in this text have a firm grasp of, albeit from differing perspectives. The issues that compelled Paula Scher to be direct and creatively aggressive in presenting her solutions for the Citibank identity during the upheaval that comes with mergers between two corporate entities with divergent cultures, while specific to that project, were driven by her unique personality and straightforward manner. She worked with clients to achieve a desired end that added to her position as one of the most successful and relevant designers today. She has extraordinary vision. As I sit writing the epilogue that you are reading, from Mi Qiu's cavernous studio on the east side of Shanghai's famous Bund overlooking the Huangpu River, I can make out the Citigroup logo in the distance atop the beautiful skyscraper that is the company's headquarters in the Pudong area of Shanghai. Just minutes away, situated right on the Bund's main road is a Citibank branch where the original logo created by Scher greets customers from the streets as they walk the busy sidewalks of Shanghai. The Citigroup logo, while adopting the font style of the Citibank identity, essentially attempted to reinforce the brand's essence, painstakingly developed by Scher. The logo utilizes the old Travelers' red umbrella attached to the end of the logotype, just as it was originally used with the Travelers' serif-faced logotype. That original umbrella was the basis for the red arch graphic atop the Citibank mark created by Scher. The Travelers Company is no longer a part Citigroup, and has subsequently merged with the St. Paul Companies, a competitor, forming the insurance conglomerate the St. Paul Travelers Company. As noted recently in a *New York Times* article on June 20, 2006, by Eric Dash: "The red, familiar and ubiquitous may finally be on its way out" as the company painstakingly considers discontinuing its use of the umbrella. I must admit to being troubled by the continued use of the red umbrella as it was clearly associated with the Travelers' insurance brand in my mind and had no definable connection to Citigroup particularly since that part of the business no longer existed.

As noted in the chapter that highlights Scher's amazing ability and direct style as a designer (who has no connection to these recent graphic developments), we should not attempt to find the rational behind Citigroup's use of the umbrella. It is just another example of the sometimes fickle nature of the relationship of the ever-changing corporate structure of big business—and design.

Lyndon B. Johnson's noble but failed attempt in creating his Great Society can be a model for the opportunity that design has to be an agent of change in the creation of the next great society. President Johnson's goal was to create a society that would support the people in their quest—human rights to have the opportunity to live, work, and play—without bias and restrictions put on them based on class, so-called race, and culture. Like any complex living organism, which all societies are, design has and can continue to be an important factor in the human experience and further development of said society. Being as concerned with the impact that products and services created and designed will have on the people who will use them must always be part of the process. These questions must always be asked: Who is this product for? How will it improve their quality of life? Do they need it?

Yes, we can do better.

Altruism as Design Methodology

It should be noted that in a recent MIT Press publication of *Design Issues* (Spring 2005), the paper "Altruism as Design Methodology," David Stairs looked provocatively at the alternative to traditional for-profit design practice, which you can easily argue fuels our present culture of consumption overdrive. Design for the World, Design for Social Impact, and Designers without Borders are firms that take a noncompetitive approach to design practice—an approach that is community based and will perhaps become more pervasive in its application in emerging countries and developing societies.

Citing Emerson, Kant, and others, the author's summary provides a sobering quote by David Korten, reminding the reader that "in a corporate libertarian dominated system, altruism is not considered good business," which brings us back to the paradox. Somehow, a balance can be found. As in the case of President Johnson's Great Society, out of the many failures attached to his vision, there were some victories. It is a worthy and important mindset—that of the design of society—one that all design practitioners should aspire to on some level.

Selected Bibliography

Codrington, Andrea. *Kyle Cooper*. Yale University Press, New Haven, Connecticut: 2003.

Cooper, Kyle. *Prologue Website*. http://www.prologue.com (6/2006)

Corporate Design Foundation. *@issue: Getting a Grip on Kitchen Tools*. Volume 2, No. 1. http://www.cdf.org/journal/0201_oxo.php (6/2006)

Corporate Design Foundation. *@issue: Stanford's Sarah Little Turnbull on Design*. Volume 7A, No. 1. http://www.cdf.org/journal/0701_turnbull.php

Corporate Design Foundation. *@issue: Kellogg Dean Dipak Jain on Design*. Volume 8, No. 2. http://www.cdf.org/journal/0802_dipakjain.php (6/2006)

Hall, Peter. *Sagmeister: Another Self Indulgent Design Book*. Booth-Clibborn Editions, London: 2001.

Hall, Peter. *Tibor*. Princeton Architectural Press, New York, NY: 2000.

Hollis, Richard. *Swiss Graphic Design: The Origins and Growth of an International Style*. Yale University Press, New Haven, Connecticut: 2006.

Henrion, FHK and Parkin, Alan: *Design Coordination and the Corporate Image*. Studio Visat London Reinhold Publishing, London 1967.

Nunoo-Quarcoo, Franc. *Paul Rand: Modernist Design*. The Center for Art & Visual Culture, University of Maryland, D.A.P. Distributed Art Publishers, New York, NY 2003.

Poynor, Rick. *No More Rules: Graphic Design and Postmodernism*. Yale University Press, New Haven, Connecticut 2003.

Scher, Paula. *Make It Bigger: Paula Scher*. Princeton Architectural Press, New York, NY 2002.

Schleger, Pat. *Zero: Hans Schleger—A Life of Design*. Princeton Architectural Press, New York, NY 2001.

Sutner, Ladislav. *Prague New York Design in Action*. Museum of Decorative Arts in Prague & Argo Publishers, Prague 2003.

Index